An Anthropology of Landscape

An Anthropology of Landscape

The Extraordinary in the Ordinary

Christopher Tilley & Kate Cameron-Daum

First published in 2017 by
UCL Press
University College London
Gower Street
London WC1E 6BT

Available to download free: www.ucl.ac.uk/ucl-press

Text © Christopher Tilley and Kate Cameron-Daum, 2017
Images © Authors and copyright holders named in captions, 2017

A CIP catalogue record for this book is available
from The British Library.

This book is published under a Creative Common 4.0 International license
(CC BY 4.0). This license allows you to share, copy, distribute and transmit the work;
to adapt the work and to make commercial use of the work providing attribution is
made to the authors (but not in any way that suggests that they endorse you or your
use of the work). Attribution should include the following information:

Christopher Tilley and Kate Cameron-Daum, *An Anthropology of Landscape*.
London, UCL Press, 2017. https://doi.org/10.14324/111. 9781911307433

Further details about CC BY licenses are available at http://creativecommons.org/licenses/

ISBN: 978-1-911307-45-7 (Hbk.)
ISBN: 978-1-911307-44-0 (Pbk.)
ISBN: 978-1-911307-43-3 (PDF)
ISBN: 978-1-911307-46-4 (epub)
ISBN: 978-1-911307-48-8 (mobi)
ISBN: 978-1-911307-47-1 (html)
DOI: https://doi.org/10.14324/111.9781911307433

To the memory of Tor, an extraordinary Border Collie who knew the heath better than either of us.

Preface

The research for this book took place from 2008 to 2012. It ran in tandem with an archaeological project involving a fieldwork survey of the entire heathlands, and excavations of multiple sites during the same time period directed by Chris Tilley. It is important to acknowledge this in terms of the discussion of this being a contested landscape. After moving to the area and having decided to visit all the prehistoric cairns, Chris went walking on the heathlands with Tor, his dog. These became walks with a purpose. After seeing all the cairns he decided to walk between them in order to study their relationship to each other and the unique Pebblebed landscape in which they are situated. He quickly became fascinated with the pebbles and how these bright and rounded objects transform what otherwise might appear, to the casual observer, to be a quite monotonous landscape. Realizing that this was unlikely to be just a contemporary appreciation, he then initiated the project. From an archaeologist's perspective he is trying to create a story of the past in the present: a story involving the topography; a story involving the pebbles, the land, the sea, the sky, the sun; and integrate these things into some kind of sense of how it might have been, all the time trying to link past and present. And so the anthropological project investigating the meaning and significance of the contemporary heathland and its pebbles arose.

All the research was carried out by Chris and Kate Cameron-Daum. It was very much a collaborative exercise in which both of us were engaged in participant observation and interviews with over one hundred informants. Chris and his family were living in the research area and Kate was staying with them: this had definite advantages in that the field site was quite literally entered when leaving the front door of the house. This permitted sustained engagement with both the heathland landscape and those working there or visiting it throughout the years and in all weathers and seasons. This facilitated, we believe, something of an intimate 'insider's' (the punctuation marks to be emphasized) knowledge of the landscape and the establishment of ongoing personal contacts and relationships. During the course of the archaeological research and the anthropological research discussed here the landscape has become a powerful element in the formation of our own biographies and identities.

We both wanted to study anthropology as students because we were interested in the lives of others and how an understanding of them might lead us to reflect on our own lives and experiences. Although the two cannot be separated, we did not choose to study anthropology to learn about the anthropologists conducting the research, their lives, trials and tribulations in the field. We take it as axiomatic that it is from the ethnographic self that accounts arise, that self-reflexivity in research is fundamental and that all our findings are subjective (Clifford and Marcus 1986; Clifford 1998; Okely and Calloway 1992; Davies 2008; Collins and Gallinat 2010). One of the great strengths of contemporary anthropology is that it foregrounds the subject and subjectivity rather than claiming a spurious objectivity from a supposed elimination of the self. Subjectivity forms the very basis of our knowledge of the field arising from being there, observing, talking, reflecting.

However, so-called 'auto-ethnography', foregrounding the researcher in the research, we believe has an unfortunate tendency to rapidly turn into a form of narcissistic navel-gazing in which the anthropologist, rather than the people he or she wishes to understand, takes centre stage. Taken to its logical extreme, anthropology then becomes a discipline that is about itself and the personalities and lives of those involved – who would really want to study that?

Many discussions of this conflate what to us are two rather different concepts: the personal and the subjective. While research is subjective this does not mean that the personality and life-history of the researcher and the circumstances in which the research has been undertaken have to be discussed and foregrounded as fundamental starting or ending points for analysis or alternatively as a form of constant dialogic encounter in the text. We, the researchers, are of course present throughout the text. Everything we have discovered arises from our subjective presence in the field research but we do not wish to turn the spotlight brightly on ourselves. The anthropological 'stage' belongs firmly to our informants and their lives. We have been the stage managers of the text and were present during the four-year performance of persons and groups in landscapes that we recount. Our textual presence only surfaces when absolutely necessary or in situations when we ourselves became some of the main actors, or to occasionally exemplify social practices through our own involvement in them.

The vast majority of anthropological research still follows the traditional model: the isolated anthropologist and his or her people with whom he or she interacts. We do not believe that this is a satisfactory research model to follow in the future. The research undertaken here has involved our active collaboration throughout and that, we believe, has had some positive outcomes. We will mention here a few of them. Our different genders meant that if people were reluctant or uncomfortable talking to one of us they might do so to the other. This was particularly important in the context of the male culture of the Royal Marines and so, while both of us conducted interviews, it was only Chris who camped out with them during their training exercises. Our differences in 'seniority' (a professor and an independent researcher, known to some of our informants) also made a difference in that a few people who in a number of cases acted as 'gatekeepers' to meeting others only felt it worthwhile engaging with someone who was, in their perception, important. Conversely this was off-putting to others who felt much happier talking to Kate.

By undertaking multiple interviews with some people, usually with anything up to six or twelve months in between, we were able to discuss between ourselves what we had learnt and attempt to address obvious failures in the kinds of information and insights that we had acquired, or information that was contradictory or ambiguous at best. But most importantly we were able to support each other and discuss as we went along, engage in dialogue with regard to what to do next and develop a further interpretative understanding. The outcome of anthropology is not a research result but a form of conversation with others, and a conversation is not about results but an end in itself. Having engaged in a long dialogue between ourselves, we offer the text as a way of engaging in one with others.

Chris Tilley and Kate Cameron-Daum, May 2016.

Acknowledgements

Heathland managers

We are indebted to Pete Gotham, Bungy Williams, Toby Taylor, Tom Sunderland and John Varley for their time and patience and for multiple interviews and walks. Bungy Williams in particular was of enormous help to us for his wealth of personal knowledge and for suggesting people to whom we might talk.

Royal Marines

We are most grateful to all members of the RM who have talked to us and from whom we have learnt so much. These include recruits Daniel Chapman, Peter Coleman, Paul Johnson, Lee Page, Stuart Palmer, Jo Saunders and James Whittet; Lieutenant Colonel Steve Wilson, Ministry of Defence Estates, Woodbury Common; George Green, Range Manager, Straight Point; Major Chris Fergusson, Commando Training Centre, Lympstone; at 121 Troop Sergeant Tim Hughes, Corporals Aaron Cox, Jack Faulkner and Alistair Stubbs and Troop Commander Nicholas Broadbent; at 122 Troop Sergeant Jim Burston, Corporals Sean Gascoigne and Jonathan Talbot and Captain Russ Sayer.

Chris Tilley would like in particular to thank Lt Colonel Steve Wilson for arranging a visit to the Straight Point firing range to interview officers and recruits in October 2010, and most especially Major Chris Fergusson for arranging field visits and accompanying him to 121 Troop and 122 Troop on Woodbury Common in November 2010. Thanks to 121 and 122 Troops for their hospitality and for allowing me to camp out with them and observe all aspects of basic training exercises. Photographs were not permitted and a draft copy of the text of Chapter 3 was vetted by personnel at the Royal Marines Commando Training Centre, Lympstone and approved for publication.

Environmentalists

Interviews were conducted between 5 April 2009 and 11 August 2009 in a variety of locations, including Aylesbeare and Venn Ottery Commons, people's homes and a café in Colaton Raleigh. We are most grateful to the environmentalists for giving of their time, thoughts and ideas and allowing us to take part in volunteering sessions. Our particular thanks go to Brian, Bonnie Blackwell, Richard Halstead and Louise Woolley.

Aggregate Industries

Our thanks to Jerry Foxall at Black Hill quarry.

Cyclists

Interviews were conducted between 10 August 2009 and 10 December 2011 in a variety of locations including places of work, homes and cafés in the Woodbury and Exeter area. We are very grateful to these cyclists for giving us their thoughts and time, in particular to Stuart Brooking, Sam Cann, James Ephraums, Kirby James, Colin and Chris at Knobblies Bike Shop and Sarah Skinner. We also would like to give our thanks and appreciation to two very special cyclists who have sadly passed away. To Kimmo Evans, Community Development Officer with the East Devon Area of Outstanding Natural Beauty and a keen member of the Axe Valley Pedallers, who organized family friendly rides during Heath Week, and to Paul Goffron, who was so very helpful with his ideas for routes and further contacts.

Horse riders

We are indebted to staff at Dalditch stables and independent horse riders, Rose Jesson, Karen Williams and Jackie Cox for talking to us about their horse riding experiences at home and on the Commons.

Walkers

Interviews were conducted between 25 May and 1 September 2009. We are extremely grateful to the walkers for giving us their time and sharing their experiences of walking the heathland. We are particularly grateful to Jim Cobley, Michael Downes, Sally Elliott, Caroline and David Keep, Stuart Lovett, Roger Stokes, Margaret Wilson and members of the Otter Valley Association.

Artists

Interviews with artists practising in a variety of mediums were conducted between 27 May and 14 August 2009. We are most grateful to them for their time and in particular to Jon Croose, Barbara Farley, Barbara Hearn, Debbie Mitchell, Caroline Saunders, Priscilla Trenchard and Michelle Wilkinson.

Fishermen

Interviews were conducted at Squabmoor in June, October and November 2010. Many thanks to Anthony Locke, Geoff Vincent, Fisherman Bowie and five other fishermen who talked to us and shared their knowledge.

Model aircraft flyers

Interviews were conducted in July 2009 in a club member's home and in October 2011 at the airstrip. Our grateful thanks to the flyers, especially Mike Bramblehay, Mike Jones and Felix Marten, who so enthusiastically shared their memories, knowledge and experience and provided us with a wonderful display.

    We are grateful also to Jim Cobley for assisting us in the car park survey. Chris is indebted to Jon Hanna for taking him up across the heath to take aerial photographs. We would like to thank Chris Penfold at UCL Press for his support for the book and advice, Laura Morley for her expertise in copy-editing and Sarah Rendell at OOH for her help during the production of the book. Last, but not least, we are most grateful for comments by two anonymous reviewers whose suggestions helped us to improve the text and provoked further reflection.

Contents

List of figures	xvi
List of tables	xix

1 The anthropology of landscape: materiality, embodiment, contestation and emotion — 1

Part I: The heathland as taskscape — 23

2 Managing the Pebblebed heathlands — 25

3 Bushes that move: the Royal Marines — 84

4 Environmentalists: the giving and the taking away — 125

5 Quarrying pebbles — 152

Part II: The landscape as leisurescape — 163

6 Introduction: the public and the heathland — 165

7 Modes of movement through the landscape: cycling and horse riding — 175

8 The cry of the Commons: walking through furze — 213

9 Art in and from the landscape — 234

10 Fishing and the watery pursuit of 'pets' — 262

11 Model aircraft flyers: spirals and loops in the sky — 273

12 Conclusions — 287

References — 300
Index — 321

List of figures

1.1	The location of the East Devon Pebblebed heathlands	15
1.2	The main places on the heathlands discussed in the text	16
2.1	Wet heath, Aylesbeare Common	34
2.2	Dry heath, Aylesbeare Common	34
2.3	Cherry trees	35
2.4	Gas Pond	37
2.5	Place names on the Aylesbeare RSPB reserve	38
2.6	'Savannah' heathland, Woodbury Common	39
2.7	Toby Taylor's map	40
2.8	Bungy Williams' map	41
2.9	Vegetation scars caused by pit digging by the Royal Marines, Colaton Raleigh Common	49
2.10	Heathland management mosaic: Aylesbeare and Harpford Commons	51
2.11A	Harpford Common: swaled areas (light circular areas in background)	52
B	Heathland management on Woodbury Common: mature heath (left), newly cut heath (right)	52
C	'Snake bend' firebreak with path, Colaton Raleigh Common	53
D	Artificial pond, Colaton Raleigh Common	53
2.12	Eighteenth-century landscaping mound, Woodbury Common, Four Firs	55
2.13	Tom's map (Natural England advisor)	57
2.14	Topsoil-scraped area, Aylesbeare Common, with vegetation regrowing in basal ditch sections of circular structures	59
2.15	The effects on the heathland of a summer wildfire, 2010, Colaton Raleigh Common	65
2.16	Woodbury Common in the 1930s. Photograph by George Carter	67

2.17A	Tall gorse bordering footpath	79
B	Gorse close up	79
3.1	A Royal Marine sheeptrack leading up to a Bronze Age cairn used as an orientation point	88
3.2	The grenade range	88
3.3	Recruit's map 1	96
3.4	Recruit's map 2	97
3.5	RM Lt General's map	98
3.6	Trainer's map	99
3.7	A harbour area	101
3.8	The endurance course	106
3.9A	Exit of dry tunnel	108
B	Gully	108
C	Peter's pool	109
D	Pebble path	111
E	The sheep dip	111
F	Crocodile pit	113
4.1	Environmental volunteer's map 1	146
4.2	Environmental volunteer's map 2	147
5.1	Part of the Black Hill quarry, aerial view	154
5.2	Sand tip and water-filled pebble extraction hole, Black Hill quarry	154
7.1	Riding group out on the heathlands	183
7.2	Riding group	183
7.3	Bike Bus T-shirt	186
7.4	Map of heathland cyclist	189
7.5	Group of horse riders on the heathlands	202
7.6	Horse-riding partners	202
7.7	Karen's map	205
7.8	Jackie's map	206
8.1	Walker's map 1	230
8.2	Walker's map 2	231
9.1	Margaret Dean's painting of the heathlands hanging in the RM Officers' Mess at Lympstone Commando Training Centre	242
9.2	Pebble painting 1	252
9.3	Pebble painting 2	252
9.4	Woven Flame 1	257
9.4A	Woven Flame 2	257
9.5	[Trans]figure monoprint	258
9.6	Pebble grid	260

9.7	Pebble prints on cotton	260
10.1	Squabmoor reservoir looking north	263
10.2	Sign on fishing swim	265
10.3	Geoff's map of Squabmoor	266
10.4	Carp fisherman at Squabmoor	272
11.1	A model aircraft enthusiast and his plane	278
11.2	The humanized cockpit	281
11.3	Model aircraft flyer's map	283
12.1	Pebble memorial to the archaeologist George Carter	292
12.2	Memory collage	299

List of tables

1.1	Informants interviewed, by category, age and gender	17
2.1	Proportions of different heathland types	45
6.1	Frequency of visits to the Pebblebed heathlands	166
6.2	Other commons mentioned by the informants	167
6.3	Likes and dislikes of visitors to the Pebblebed heathlands	169
6.4	Respondents' knowledge about the presence of endangered species on the heathlands	171

1
The anthropology of landscape: materiality, embodiment, contestation and emotion

Introduction

Landscape is a subject of study that belongs to nobody. It has long been studied in various ways and under various guises by geologists, social and cultural geographers, planners, ecologists, historians and art historians, archaeologists and anthropologists. Landscapes form the basis for much poetry and innumerable novels and are thus of interest to literary critics. Discussions of landscape are a mainstay of much social and political journalism. To be interested in landscape is thus to enter a promiscuous field criss-crossed by different theoretical and methodological perspectives, values and interests. To some this undoubtedly makes the topic exasperating; nobody can adequately define or tie down the term, it is out of control and therefore of no analytical value. To others, such as ourselves, the inherent ambiguity of the term and the diversity of approaches and perspectives used to study it is precisely that which makes the study of landscape so interesting and valuable. Such a topic is inexhaustible and unbounded; rhizomic rather than rooted (Deleuze and Guattari 1988: 5–25), perspectives on landscape pop up anywhere and often in an unpredictable manner. In many of these studies the term never appears because others such as space and place and the environment – even more broadly, the world – subsume it.

Landscape is thus an absent presence in a huge body of scholarship. In anthropology, books with landscape in the title were virtually absent twenty-five years ago (Tilley 1994). Since then there has been a growing interest in and development of landscape studies in books (Bender 1993; Hirsch and O'Hanlon 1995; Feld and Basso 1996; Ingold 2000; Bender and Winer 2001; Stewart and Strathern 2003; Tilley 2006; Arnason *et al.* 2012; Jarowski and Ingold 2012) and in many journal articles. While the traditional output of research in social and cultural anthropology has been the ethnographic monograph hardly a single one has appeared foregrounding the study of landscape as a topic worthy of consideration in its own right during the last two decades. Ethnographic studies of landscape are thus usually compressed into small vignettes within an overall disciplinary field that swallows them up. An exception can be found in the recent studies of Laviolette (2011a; 2011b). One of these volumes is about landscape only in a metaphorical sense, its focus actually being on extreme sports such as cliff jumping, extreme surfing and urban parkour. The other considers a huge region, Cornwall in south-west England, from a variety of different perspectives, with its chief focus being how cultural metaphors of identity are materialized. In its consideration of a variety of different social groups – amateur footballers, artists, farmers, fisherfolk, immigrants, landscape gardeners, scholars and tourists – it comes closest to the general perspective taken up in this volume. But Laviolette's landscape analysis is on a macro scale. It embraces a whole series of different landscapes within Cornwall, like a series of Chinese boxes, one inside the other. His informants, by and large, don't bump into each other in their daily lives as they are dispersed over a huge peninsula. This study by contrast considers a small-scale landscape from different individual and social perspectives, enabling us to consider embodiment, materiality and contestation in a quite different manner because our informants are constantly co-present with others in the same landscape.

This book is an extended study of a particular rural landscape in south-west England. While we have no wish to rigidly define the term landscape we want to briefly highlight below what we regard as the main features of this particular landscape study and what it may have to offer.

- Biography: we examine the biographies of persons and the manner in which the landscape becomes part of whom they are, what they do and how they feel.
- Place: we discuss the manner in which different individuals are involved in place-making activities, that is to say how they name places, sometimes not places on any Ordnance Survey topographic map, the places they like or dislike (Tuan's topophilia and topophobia; Tuan 1974, 1977). In this respect we consider landscape as being a set of relationships between places in which meaning is grounded in existential consciousness, event, history and association: wisdom 'sits in places' (Basso 1996).

- Motility: we discuss the manner in which persons and groups move across the heathland landscape: the paths that they follow and the manner in which they move, on their own or accompanied by others. The temporality of movement and the sequences in which persons encounter places along the way may be fundamental to how people experience landscapes and thus feel about them (Tilley 1994: 27ff.; Ingold 2007, 2011).
- Mediation: we discuss how the manner in which the heathland is encountered and understood alters according to whether people walk across it (and the manner in which they walk; Ingold and Vergunst 2008; Tilley 2012) or whether their encounter is technologically mediated – by modes of transport such as cycling; by activities involving tools such as fishing, flying model aircraft or holding a rifle; by riding across it on a horse; or by being accompanied by a dog.
- Agency, aesthetics, and well-being: we consider what the landscape, as a sensuously encountered material form, does for people and in reciprocal relationship what it does for them (Gell 1998; Milton 2002; Tilley 2004, 2008, 2010; Laviolette 2011a).
- Conflict and contestation: we discuss the ways in which differing attitudes and values to landscape relate to different modes of encounter and priorities: the politics of landscape (Bender and Winer 2001; Tilley 2006).
- Nature and culture: what do these terms mean to people in the context of this landscape? While academics happily dispute the value of the opposition (e.g. Descola and Palsson 1996; Descola 2013; Darrier 1999; Strang 1997; Ingold 2000; Castree and Braun 2001; MacNaughton and Urry 1998), nature is to others an invaluable term informing their environmental ethics and politics and their encounters with the world. To strip a concept of nature away may thus have unintended and disempowering social and political effects in terms of a rapidly developing global crisis in which humanity is destroying the environment on which it depends.

We consider the archaeological and historical development of this landscape in a companion volume to this. Anthropology rapidly turns into history. In fact it is already history by the time that it is published. The ethnographic present of this book is the period 2008–2012, when the fieldwork was carried out. We wish to elaborate below in much more detail on four key concepts that inform the structure of the entire book: materiality, embodiment, contestation and emotion.

Materiality

A considerable amount of recent scholarship concerned with landscape has stripped it of its materiality. By this we mean that the research is thoroughly mediated by discourses and representations. Examples include writings, maps, photographs, paintings, drawings, an entire apparatus by means of which we vicariously inform ourselves about something out there and distant from our desks. We see and understand landscapes through the representations of others and, in turn, these representations become the object of further discourses. So in a somewhat bizarre manner cultural geographers Cosgrove and Daniels can define landscape as 'a cultural image, a way of representing things' (1988: 11). Matless (1998) discusses the English rural landscape largely in terms of its iconographic representation. Images take precedence to people and place. Other scholars similarly taking a 'post-structuralist' turn instead assimilate landscape to text. Duncan conceives landscape as 'one of the central elements in a cultural system, a text' (Duncan 1990: 17). Such a text is a signifying system through which a social system is communicated and experienced: one reads it like a book, and one does not necessarily need to be there in order to do that, to experience it; indeed one does not need to talk to anybody in order to write about it in a univocal fashion (see for example Gregory's astute comments (Gregory 1994: 298ff.) on Soja's (1989) representation of the Los Angeles urban landscape). Daniels and Rycroft (1993) are content to map modern Nottingham through the novels of Paul Sillitoe, rather than gaining knowledge through walking the streets. We are not arguing that pictorial or textual representations of landscape are uninteresting or unimportant to analyse (see e.g. Laviolette's anthropological mapping of Cornish identities in terms of images (2011b: 80ff.), nor contesting that they may constitute very powerful ways through which people know and experience physical landscapes, so much so that texts or imagery begin to constitute and structure encounters and experience of material landscapes. Quite the contrary, it is just that they have tended to dominate much discussion. Indeed, they have been taken by some as defining what landscapes actually are and what the object of a landscape study actually is. We offer a thoroughly materialist approach here as an antidote and counterpoint.

From our perspective in this book representations of landscape, textual or pictorial, are of secondary significance and we should treat them as such; they are selective and partial, and often highly ideological, ways of seeing and knowing. In fact it is through material experience that we can understand the ideological nature of these representations, the manner in which they quite literally frame the landscape, far better than by undertaking any desk-bound analysis. We make the simple and somewhat blindingly obvious comment that walking is not a text, cutting down a gorse bush is not a text, training to be a soldier is not a text, a body is not a text, hills and rivers and trees are not texts. A materialist approach to landscape is thus a return to the real, and we regard it as a way to reinvigorate and redirect the study of landscape. The move is from representation to the materially grounded messiness of everyday life and the minutiae of material practices that constitute it. A stress on the materiality of landscape means that the anthropologist/researcher needs to be there, to experience the landscape through the sensual and sensing body, through his or her corporeal body. The body becomes a primary research tool. Such an emphasis on being there and observing and interacting with others stresses performativity: the manner in which our identities and those of others are constituted in and through action, and the manner in which these identities come into being through performances of identity (Butler 1990).

Fortunately there is a very long tradition in anthropology of participant observation and subaltern studies on which to draw, one that has continued to have a very significant impact on the ways in which anthropologists have written about landscape and that is manifested in many of the various studies cited above. As Ingold has cogently noted: 'we owe our very being to the world we seek to know. In a nutshell participant observation is a way of knowing <u>from the inside</u> ... Only because we are already of the world, only because we are fellow travellers along with beings and things that command our attention, can we observe them' (Ingold 2013: 5). We also draw on another rich and increasingly prominent anthropological tradition, that of material culture studies themselves (for recent work see e.g. Tilley *et al.* 2006; Miller 1998, 2005, 2010; Ingold 2013). These involve an insistence that persons and things are mutually constitutive. A landscape is certainly a complex kind of thing. Unlike an artefact, we cannot grasp it in our hands or move it around at will. It forms a material medium in which we dwell and move and think. We are not somehow outside it, or contained by it; landscape is part of ourselves, a thing in which we move and think. Therefore we cannot think of it in any way we like. It is not a blank slate for conceptual or imaginative thought but a material form with textures and surfaces, wet and dry places, scents and sounds, diurnal and seasonal rhythms, places and paths and cultural forms and built architecture that, through differential experience, is constitutive of different identities. So the landscape is both inside the body and outside of it, both part of whom we are and a thing apart. Persons and landscapes are entangled in a network of material and social relations (for general discussions of the intertwining of persons and things and their consequences see Olsen 2010; Hicks and Beaudry 2010; Hodder 2012) providing both affordances and constraints for the performance of identities that always occur in particular material and cultural contexts. Landscape is thus an intertwining of the flesh of the body and the flesh of the world, to use Merleau-Ponty's metaphor (Merleau-Ponty 1968: 142). Landscape is undoubtedly a very complex material thing to attempt to understand or make sense of since it is, to use Latour's (1993) term, a quasi-object, something constructed and made; a cultural product, but having an independent existence with its own rhythms and purposes. We are touched by this fleshly material world of landscape and in turn touch it. In the process we transform ourselves.

Embodiment

Embodiment is a key term informing the discussions of this book in the individual chapters in Part I and II. Here we wish to briefly outline what is meant by this term from a phenomenological perspective broadly inspired by the philosophical writings of Merleau-Ponty (1962, 1968), and other interpreters of his work. Collapsing a mind/body dualism, the body is both object and subject, but the relation between the two is internal so that subjectivity does not arise in the mind or in consciousness but is in the body. Both subjectivity and the physical character of the body as a thing or object are related to the corporeality of body and mind: what a body is and what a body can do. The whole notion of a disembodied consciousness is simply a manifestation of idealist thought itself. Such a consciousness cannot exist because the mind inheres in the body and is not independent from the body. It follows that the kinds of distinctively human bodies that we have are part and parcel of the manner in which we think about and experience the world. Our consciousness is thus structured in tandem with our bodies as sensuous, carnal and subjective things.

Merleau-Ponty argues that our sensuous perceptual activity ends in objects, a position that runs counter to the naïve empiricist view that assumes a world of impressions and stimuli that exist in themselves in relation to which the body responds and reacts. Instead, the body constitutes both the cognitive ground of culture and its existential ontological ground (Lakoff and Johnson 1999; Cszordas 1990; Desjarlais and

Throop 2011; Jackson 1995, 1996). Objects are a secondary result of thought. This does not mean that these objects are immaterial or purely a product of the mind. Instead objects are part of the same social and material world that we inhabit. We 'produce' or 'recognize' them through reflecting on that world and the process is indeterminate insofar as we can never sense the entire world from the determinate situatedness of our bodies. We exist in the world and relate to it from a point of view – the setting of our bodies. So perception begins in the 'pre-objective' material and subjective body and ends in the objects that the body perceives in relation to it: 'my experience breaks forth into things and transcends itself in them, because it always comes into being within the framework of a certain setting in relation to the world which is the definition of my body' (Merleau-Ponty 1962: 303).

The bodily setting in relation to the world that we are concerned with in this book is that of landscape, which provides, we argue, an existential ground for our embodied being: we are both in it and of it, we act in relation to it, it acts in us. Landscape is a product of our reflective activity arising from our pre-reflective or pre-objective bodily relation to it (for a detailed discussion see Marratto 2012). Bodies and landscapes thus produce each other in mutual relation, in the process of motility and inhabitation. In the most basic sense the agency of landscape is embodied because it acts on us through the mediation of our bodies. The thinking, subjective mind emerges in relation to the landscape and ends in its perception. Thus the body may be both subject and object, sensing and sensed within a landscape setting. It may be experienced from the 'inside', through kinaesthetic sensations conveying information about posture, position and movement, or from the 'outside' as a body among others intersubjectively constituted through a mutual relation with other persons in culture.

A seemingly contrasting perspective is provided by Latour (2004), who argues that the body should instead be conceived as an interface between different subjectivities and objects; it is from this that perception arises. He makes no distinction here between 'natural' objects and material culture objects. Both play an equally important role in the constitution of subjectivity rather than being a product of bodily perception that cannot exist anterior to perception. This is a perspective used by Vilaça (2009) in a discussion of Amazonian bodies used to critique an 'embodiment paradigm'. What is at stake here is exactly how we regard the primary locus of perceptual activity taking place, and it seems to us a kind of chicken-or-egg question lacking any satisfactory answer.

Lakoff and Johnson (1990, 1999) explore the manner in which our everyday cognitive capacities are rooted in relation to our bodily being and emotional capacities in contemporary western culture: the manner in which we perceive things to be near or far, to the left or right of us, behind or in front of us, below or above us, forms the basis for our everyday, ordinary taken-for-granted and pre-reflective metaphors by means of which we represent the real in language: the foot or brow of the hill, the face of the clock, the legs of the table and so on. Happy is up, sad is down, etc. etc. (1999: 49ff.). Metaphors are an ever-present part of our language and the way in which we represent the world. They form particular understandings of the landscapes we inhabit and the manner in which they are empowered or naturalized (Tilley 1999, 2004).

Lakoff and Johnson point out strongly that because reason is not independent of perception and emotion the distinction between animals and humans is not easily drawn. In fact human reason is a form of animal reason (Lakoff and Johnson 1999: 17) because both have a bodily basis involving categorization of food, mates, predators and members of the same species. Such reasoning obviously differs nonetheless in terms of the manner in which it is embodied and through the perceptual senses. Human conceptual reason does not reflect external reality because it is mediated and shaped by the sensorimotor capabilities of our bodies, as it is for other animals. This is important in understanding the embodied relations between persons and animals and the manner in which each understands and perceives the other, so much so that we may consider persons and animals in some instances, such as the rider on a horse, or a dog and a dog-walker, as co-beings mediating each other's relationship to the landscape (see discussions in Chapters 7 and 8).

While animals may actively mediate a human embodied relationship with the landscape, so do technologies. In our everyday pre-reflective relations with the world we do not think in terms of subject–object relations. We typically use tools as extensions of our body: the soldier and his rifle (see Chapter 3) or the fisherman (gender intended) and his rod and line (see Chapter 10). We become adept at using them and only atypically do we regard them as objects of contemplation. Much work on technology, while elaborating on the processes of making and using things and describing them in terms of complex operational chains, has tended to neglect consideration of their sensuous embodied material character. Things extend our sensorimotor capacities out from the body and into the setting of the world. In the process perception and understanding may be materially extended. In this case the agency of things consists in their ability to shape and mediate human actions. They do this as part of a field of relations with others, a domain of social practice, a dialectic of embodiment and objectification or a bringing forth into the world (Bourdieu 1977: 87ff.; Tilley 2006a). Warnier's 'praxeological' approach (2001) usefully fuses a consideration of bodily techniques and instrumental techniques to understand how skilled practices become subjectively internalized (for a variety of perspectives on these themes see Ihde 1990, 2002; Ingold 2013; Lemonnier 2013; and most especially Coupaye 2013).

Another key aspect of an approach emphasizing embodiment is a consideration of spatio-temporal relationships. Space and time are not somehow outside social relations and acting to contain them but arise from their embodied relation to persons. So what is near or far, here or there, bounded or unbounded differs in relation to the body itself and its motility in the world. So duration and the 'depth' of the landscape and what constitutes the horizon become part of the pre-objective constitution of bodily perception. Past experiences feed into the present, anticipating the future. Our temporal experience 'colours' the manner in which we understand the present from the lived perspective of the body. This is always limited, ambiguous, shifting and changing; some aspects of landscape become foregrounded at one temporal moment and fall into the background at another. Embodied perception shifts and changes, is always in flux and is related to our interactions with sentient others, human or non-human. Our perceptual senses engage with our embodied being all at once in synaesthetic relation. We do not see the world and then hear it or smell it or touch it. All our perceptual senses intermingle in our embodied experience and all at once, a position currently being valuably explored in the emerging sub-discipline of sensory anthropology (Classen 1993, 2005; Stoller 1989; Howes 1991, 2005; Pink 2009; Ingold 2011).

Contestation

Meinig (1979) invites us to imagine a landscape thus: a group of different people go to the top of a hill and look down and across the panorama of landscape below. Each is invited to describe the landscape before them: what do they see? Meinig lists ten versions of the same scene: the landscape may be regarded in various ways as nature, habitat, artefact, system, a problem, as a source of wealth, as ideology, history and so on. Why the people might describe it in these very different ways relates to their point of view and their interests and values, so inevitably the landscape seen from the 'beholding eye' means something radically different for a property developer, a local historian, an earth scientist, an artist and so on. Ten versions of the same thing is obviously an arbitrary number: there could be many more or less. The general point though is that political, economic, moral and aesthetic interests and values colour what people see and may inevitably lead to radically different attitudes.

Landscapes are thus inevitably contested. They are valued precisely because they are valuable, part of people's lives. They reflect the complexity of their lives. They are historically contingent and their mutability stems from the various ways in which people understand them and engage with the material world. So landscapes are untidy and messy, tensioned, always in the making (Bender 1993, 1998, 2006; Bender and Winer 2001). Our landscapes of modernity are frequently on the move and peopled by diasporas and migrants of identity, people making homes in new places. They may be structures of feeling, outcomes of social practice, products of colonial and post-colonial identities and the western gaze, bound up with class divisions, property and ownership, outcomes of the contemplative sublime or places of terror, exile and slavery (Tilley 2006: 8). For some, an increasingly small minority, landscapes are 'taskscapes' (Ingold 2000: 189ff.) in which they earn a living. For the vast majority landscapes have become pleasure grounds where they pursue their interests and foster their own personal development. This inevitably produces conflicts of purpose and value, discussed at length in Parts I and II of this book. The landscape provides different possibilities and potentialities for different groups and that which is good for one is not necessarily so for another. Some may want the landscape to stay the same and conserve it, others may want to develop, alter or enhance it.

Emotion

Emotion or feeling resides at the heart of our human capacity to experience landscapes as meaningful and a wish to prevent their destruction. Yet, as Johnson has remarked, there is very little sustained analysis of emotional meaning in philosophy or the social sciences more generally. What is deemed subjective, private and personal is no doubt regarded as lacking any cognitive significance and such irrational responses are not seen to merit 'serious' rational discussion (Johnson 2007: 53). Referring to recent research in neuroscience Johnson points out that basic emotions such as doubt, shame, fear and joy have a deep-seated bodily basis; they may arise from the body in a particular situation in the world (e.g. seeing a venomous snake in front of you or the joy involved in the birth of a child). This pre-objective response of the body then gives rise to reflection, a process in which cultural meanings are integral. So we only realize that we feel something after a deep-seated response has taken part in embodied experience. Environmental psychologists have not shied away from considering the importance of emotional responses to environment and landscape. They have long held the view that emotion plays a key role (e.g. for early work see Kaplan 1973; Ittelson 1973; Wohlwill 1976).

Milton (2002), in a path-breaking anthropological study, has explored in depth the relationship between emotion and rationality in environmental policies and practices, in which contestation inevitably becomes an issue in relation to economic development. She examines the manner in which thoughts and feelings, goals, values and emotions emerge from personal engagement with the world. Why is humanity rapidly destroying the world? Why do we not care sufficiently about 'nature' to stop destroying it? Emotion, she argues, is the primary reason some people care about nature. Environmental campaigners are passionate about what they do and will speak about their feelings for and enjoyment of the natural world. These deep feelings for nature emerge from their perceptual experience of their environment. Working primarily with environmentalists in Britain and Ireland she astutely examines how environmental policies and practices get formulated in terms of a wider field of social and political relations. A fundamental difference between the manner in which modern western societies and indigenous traditional societies treat nature often involves the notion of the sacred. The former can destroy nature because they are separated from it whereas for the latter nature inheres in social being. Nature for us in the contemporary west is a resource to be used and exploited and bound up with land ownership.

In indigenous traditional society nature is usually not owned by individuals. It cannot be bought, sold and bounded but is sacred and intimately related to social identity. In destroying it people destroy part of themselves. Much has been written about this through the prism of the relationship of Australian aboriginal populations and their landscapes (Munn 1973; Strang 1977; Morphy 1993) to the indigenous cultures of North America in which the landscape is animate and peopled with spirits (e.g. Brody 1982; Nelson 1983; Tanner 1979; Hornborg and Kurkiala 1998). A rational scientific approach to nature has served capitalism very well by depersonalizing nature and in the process removing the moral responsibility for destroying it (Milton 2002: 40ff.). One of the cases she examines is that of a proposed super-quarry on the island of Harris in the Outer Hebrides. In terms of a rationalist western discourse on environmental issues it is useless for protestors to claim that such a development should not go ahead because they love nature and this quarry would destroy a sacred mountain. Notions of sacredness are fine to take seriously in relation to traditional indigenous societies but not amongst ourselves in the west. Instead discourses of environmental protection have to be framed in terms of a rationalist logic of cost–benefit analysis, preferably in terms of that which can be measured, such as the visibility of the quarry in the landscape and its economic impact on tourism in what was designated as a national scenic area. A huge protest in these terms against the quarry emerged (Milton 2002: 137ff.). As Milton points out, while notions of natural beauty are inherently subjective and in the eye of the beholder, visibility can be objectively measured and calculated: 'the defence of natural beauty, and the defence of the market interests that threaten it, have to be presented in an idiom that enables decision makers to appear independent. In western cultures, that idiom is scientific' (Milton 2002: 139). Milton cogently argues that the opposition between rationality and emotion is a false one. Indeed it is irrational to reject emotion as a way of relating to and valuing landscape in public policy and other decision-making processes, for emotions are what make us human.

Carrier (2003) makes a similar argument in relation to conflict over environmental conservation in Jamaica. Here, as elsewhere, the motivation on the part of conservationists arises from their personal biographies, which stimulate their desire to protect the natural world from destruction – from their emotional attachment to and knowledge of a place. However conservation policy has to be formulated in a supposedly impartial rationalist logic for it to be an acceptable discourse and to be taken seriously.

A powerful emotional attachment to a certain place may also result in a tenacious feeling of ecological identity. Zavestoski notes through interviews and observation that:

> it became apparent that most of the participants had either special places in nature, a place that had been special to them but was developed or destroyed, or a particular experience in nature that was significant in developing their concern for nature ... [and they] explained how expressing their concern for these special places as an emotional attachment or sense of oneness often resulted in strange looks or dismissive reactions
>
> (Zavestoski 2003: 304)

Again we see here the regard with which the opposition between the supposedly rational and the emotional is held, so it is refreshing to find Alfred Wainwright's exclamation 'Lakeland is an emotion' (2003: 203); for him there was no fear of being held guilty of unscientific anthropomorphism.

De Nardi's research and writings on the Italian resistance in WWII provide a rich example of how emotion is embodied in this particular contested landscape. She explicitly probes the veteran's embodied experience, focusing on

> the dynamics between space, the body and emotion, starting from the premise that collecting wartime histories means dealing with tales of the body and remembered corporeal experience (such as the discomfort and soiling of the body, and the violence perpetrated on the body) as well as through the gestures of the body in recollection ... the body is a pivotal site of memory.
>
> (De Nardi 2014: 74)

De Nardi writes that throughout the exploration of the 'worlds of feelings' of the veterans she came to appreciate the 'embodied and situated nature of much Second World War storytelling, and the paramount role played by landscape and the environment in shaping emotions, memories and approaches to the past and the events of 1943–1945' (2014a: 444).

Closely linked to the sensual dimensions of emotional experience, identity (the definition of which is of itself 'complex and contested'), memory and motility within landscape are part of well-being. Indeed, it is only since the 1990s that there has been positive recognition of the association between emotional well-being and mental and physical health and of how this is influenced by physical activity (Fox 1999: 411–418; Stewart-Brown 1998: 1608–1609). A good example of an activity recognized to have an effect on well-being is *Shinrin-yoku* – forest-air bathing – during which participants walk and breathe in the 'volatile substances' released by trees. A popular form of relaxation in Japan, it has been shown to be of great benefit in reducing high stress levels, depression and hostility, all of which are major contributors to chronic heart disease (Morita *et al.* 2007: 54–63). These studies provide an invaluable background to the consideration of the particular landscape discussed in Parts I and II of this book. People are materially entangled and entwined with landscape and precisely because of that they are emotionally bound up with its past, present and future.

The Pebblebed heathland landscape of East Devon

The material context for this study is the East Devon Pebblebed heathland in south-west England. This is an area roughly bounded by the River Otter to the east, the Exe estuary to the west, the sea to the south and the Blackdown Hills to the north. The area covered by uncultivated heathland is small. At their maximum extent the heaths cover an almost continuous area of only about 13 km north–south, and 2–3 km east to west. In places the heath is broken up by areas of improved arable land. The Pebblebed heathlands acquire their name from the distinctive geology of the area. Fringed by rich pasturelands on clays and marls, the bedrock of the heathlands is made up of multicoloured and water-worn pebbles. These are the remains of a huge river that ran through the landscape during the Triassic era some 240 million years ago. Now what once was a river bed flowing through a sandy desert is raised up to form a low ridge surrounded by farmland and, beyond that, higher hills.

This landscape made up entirely of pebbles is unique and quite extraordinary in the UK (see Tilley 2010). The area, although settled from the Neolithic onwards, has never been cultivated as the soils are very thin and acidic. Today it is largely ungrazed, consisting of an open landscape of gorse, heather and bracken criss-crossed by streams and with many boggy areas of wet heath. It is a Site of Scientific Interest and an Area of Outstanding Natural Beauty (see Chapter 2). All the historic and contemporary settlements in the area fringe the heathland; these include villages to the east and west, and the small towns of Budleigh Salterton and Exmouth to the south and south-west. The villages nestle in the valleys and it is possible to walk across the heathlands and see no trace of contemporary settlement. In some areas, furthest away from roads and car parks, one might not see a single person for an entire day except during weekends (for some historical accounts see Brighouse 1981; S. and R. Elliott 2004; Stokes 1999). Although they are small in extent they seem vast. From high points there are extensive views south off the heathlands and across the sea. The nearest city is Exeter, some 11 km to the north-west (Figures 1.1 and 1.2).

A significant part of the regional economy today is tourism, but the tourists are concentrated along the southern coastal fringe of the heathlands and very few visit or know about the heaths (see Chapter 6). Nobody dwells on the heaths today with the exception of the Royal Marines, and their dwelling is only temporary as part of their training exercises (Chapter 3). Those who work on the heaths or visit them are predominantly local people from the surrounding towns and villages and

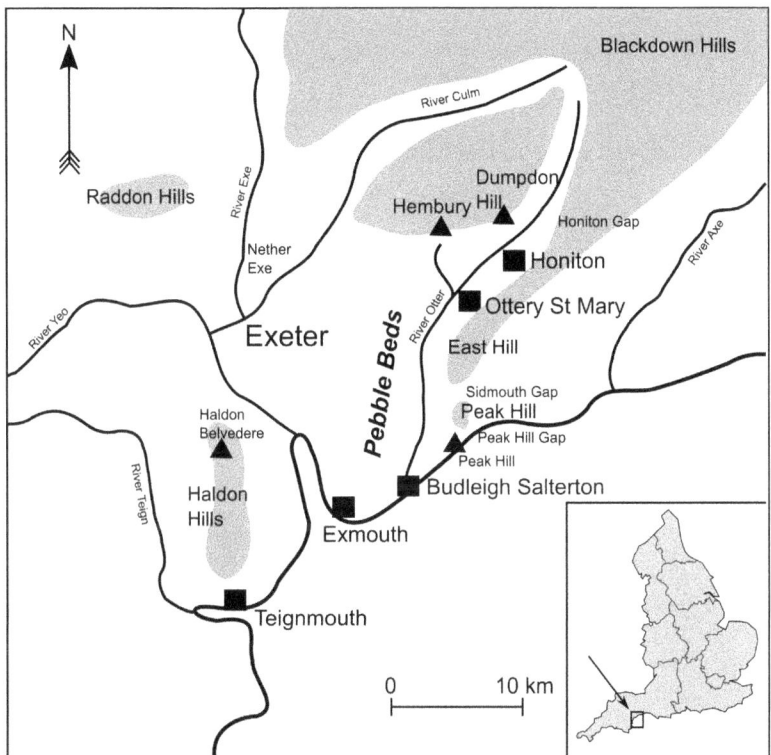

Figure 1.1 The location of the East Devon Pebblebed heathlands. Drawn by Wayne Bennett

the nearby city of Exeter. But even many locals, including people who have lived in the area all their lives, do not know anything about or visit the heathland. In this sense, it might be described as a 'secret' landscape within an otherwise quite densely settled area. In the book we consider the structure of ownership, the different groups who work in this heathland and earn a living from it, and those who use it for leisure activities. The local population is strikingly white and a significant proportion of them are elderly. Figures from the latest (2011) census show that the mean age of residents in the Budleigh Salterton area is 53.1, with 49% of people over 60, one of the highest proportions for any town in the UK, while 97% of the population is white British (Office for National Statistics 2011). Like other towns and cities along the coast of southern England it has become a favoured retirement area for wealthy outsiders, with a significant proportion of the population moving here from London and other cities in the UK.

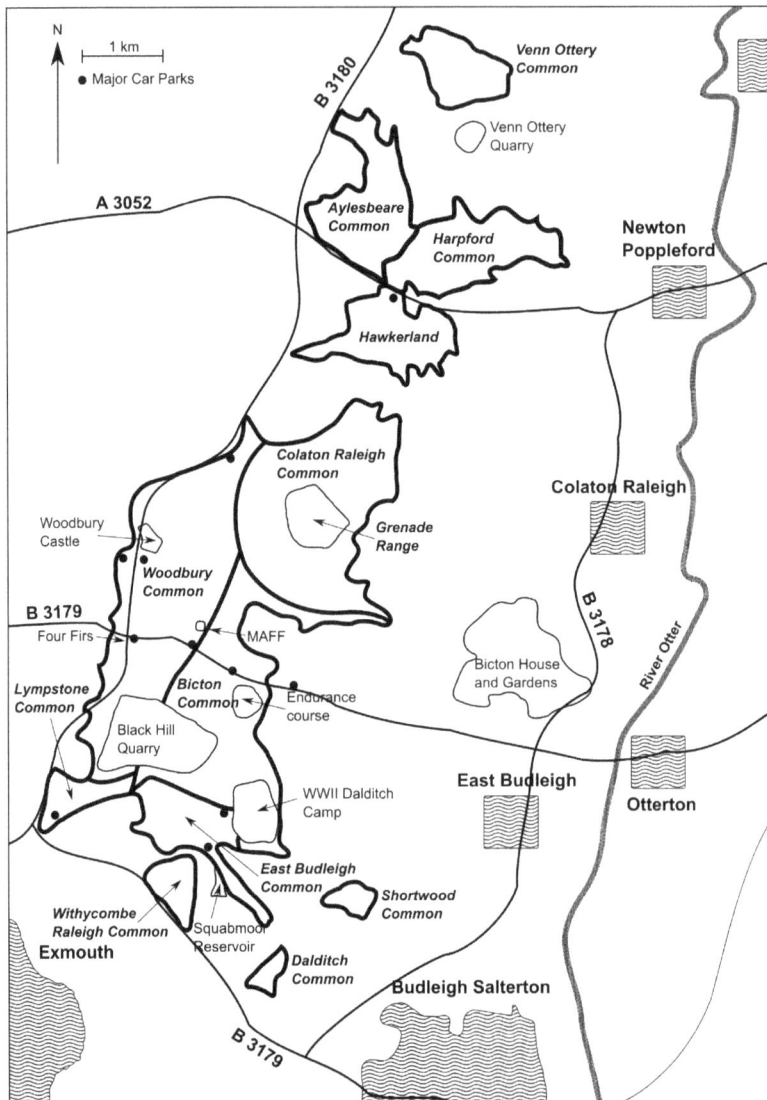

Figure 1.2 The main places on the heathlands discussed in the text. Drawn by Karolina Pauknerová

Social groups and the research field

The research is based on semi-structured interviews with 125 informants (see Table 1.1).

These were undertaken intermittently during a four-year period, 2008 to 2012. In addition to this, structured interviews were conducted in

Table 1.1 Informants interviewed, by category, age and gender. Categories are simplified insofar as eight individuals fell into two categories, e.g. as both a cyclist and a walker or a walker and an environmentalist.

Category	No. of Informants	Gender		Age 16–30	Age 31–60	Age > 60
		Female	Male			
Officials	11	4	7	0	9	2
Marines	14	0	14	8	6	0
Environmentalists	12	4	8	0	8	4
Quarry personnel and protestors	5	2	3	0	4	1
Cyclists	9	2	7	0	7	2
Horse riders	5	5	0	1	4	0
Walkers	25	8	17	0	8	17
Artists	7	5	2	0	6	1
Anglers	8	1	7	2	6	0
Model aircraft flyers	5	0	5	0	1	4
People with pebble structures	25	15	10	2	8	15
Archaeologists	7	3	4	0	6	1
	Total	49	84	13	73	47
	Percentage	39%	67%	10%	58%	37%

brief car-park surveys with fifty members of the general public visiting the heathlands (see Chapter 6). Some information concerning some of the interviewees – local people with pebble structures (i.e. walls, paths, ornamental features, etc.) in their gardens and archaeologists working temporarily on the heathlands as part of a landscape survey and excavation project – is discussed in the companion volume to this book (Tilley 2017); these discussions have informed some of the interpretation and analysis here. Six key informants, drawn from those managing the heathlands and from the Royal Marines who use them for training exercises, were interviewed on multiple occasions by both authors. The interviews took place in a variety of locations: roughly half on the heath itself, the others in people's homes, places of work and local cafés. As is conventional, all names in the text have been changed except those of persons who are too well known to be disguised. Their permission to give their real names has been sought and kindly given.

Alongside these interviews we engaged in participant observation with a variety of individuals and different groups: we walked and traversed the heathland with those responsible for managing and maintaining it, camped out with the Royal Marines on some of their basic training exercises, experiencing the landscape with them during both daytime and night-time exercises, undertook litter picks with volunteers, joined groups of volunteers engaged in environmental management, and attended public and official meetings regarding the future of heathland management and quarry development. We were involved in the annual celebration of the heath (heath week held during the last week of July) on three consecutive years. During this a whole series of events are organized for the public: guided walks and wildlife rambles at dawn and at dusk, to listen to nightjars, activities for children such as pond dipping and learning about the work of archaeologists, visits to the quarry to hear about its pebble extraction and crushing operations. One of us took guided tours to archaeological sites and monuments as part of these events. We went out walking with groups of ramblers, watched people fishing and flying model aircraft. We observed walkers, cycling groups, horse riders and the Royal Marines crossing the heath on numerous occasions throughout the years and in all seasons. Much was learnt during a systematic archaeological survey of the entire heathland landscape during this period and from the vantage points of various excavation sites during fieldwork periods. We also asked selected informants to draw for us cognitive or mental maps of the heathland (Downs and Stea 1973; Gould and White 1993). These memory maps were not a test of knowledge but were intended to provide information about place preferences, places that mattered enough to people to include them in their maps. We regard these maps as their personal representations of the heathland as being another way of telling. The heathland became during this extended period of fieldwork very much part and parcel of our own biographies and identities and we developed a deep affection and visceral knowledge of it. Much of this experience sits in bodily memory and is impossible to convey and recount in mere words. Inevitably the discussions that follow select from our experiences and what we have learnt from talking to and engaging with others. The irony of any study of embodied identities and the subjective experiences of others and ourselves is that, as a representational discourse, it attempts to write that which cannot be written: much is lost or transformed in the process.

The book can be regarded as a contribution to what has been variously labelled over the last thirty years as 'anthropology at home' or 'an anthropology of Britain' (Strathern *et al.* 1981; Cohen 1982, 1987;

Jackson 1987; Rapport 1993, 2002). So this book is about Britain as an ethnographic region of study and it is being carried out at home. However, we are not particularly comfortable with either of these labels. First of all we do not consider anthropology carried out in the nation state in which one happens to live, or have been born, to be in principle any different from research carried out elsewhere. It may, of course, be linguistically less challenging, and in purely pragmatic terms easier than conducting research in an 'exotic' location. It also obviously relieves the angst and moral burden of a discipline still tainted by colonialism and, today, by the unequal power dynamics of a post-colonial encounter with the Other.

Second, the notion that there is any such thing as British culture or an enduring sense of Britishness to be discovered and isolated in a multicultural, globalized, hybridized and creolized world 'on the move' is rather difficult to maintain (Appadurai 1996; Eriksen 2010; Hannerz 2010; Rapport and Dawson 1998). At most Britishness or British culture in any broad sense of this term is simply a manifestation of an imagined community in Anderson's (1991) sense of the term, something produced and fabricated rather than shared and lived. But this point is too blunt and requires qualification. Any notion of a British culture constituted by a coherent and integrated series of ideas, beliefs and identities shared by all contemporary British subjects does not exist in the twenty-first century and furthermore never did exist. However, there is another more humble and everyday sense of Britishness that may still be said to persist and be shared by many British subjects. This is not usually a matter of overt and conscious identity construction in flag-waving and celebrations of royalty, but it may nevertheless be objectified in a myriad of everyday *material* forms and practices such as talking about the weather (Fox 2004), gardening (Tilley 2008, 2009), pubs, the popularity of walking (Chapter 8) and coarse fishing (Chapter 10). Such different practices may be completely unrelated in the daily lives of different British subjects, who may participate in only some or none of them. Furthermore, they are cut through and refracted in multiple ways through the prisms of gender, class, socioeconomic status, ethnicity and, in terms of localities and their histories, regions and nationalism.

This book differs substantially from some mainstream cultural and social anthropology in calling into question the latter's rather narrow focus on social and political relations as the principal object of study. For example Rapport, one of the most prominent and subtle practitioners of an anthropology of Britain in his study of 'Wanet', a village somewhere in a rural dale, somewhere in north-west England, considers the landscape only as a kind of backdrop in relation to which lives and plural identities – including his own – get played out (Rapport 1993, 2010). 'Wanet' might be anywhere, or at the very least anywhere rural in this particular part of Britain. He suggests that the individual and his or her creativity is crucial for anthropological analysis, with the human psyche as central (Rapport 2002: 8). Anthropology is for Rapport thus primarily about mind, perhaps ultimately a form of individual and social psychological analysis.

The alternative view put forward here is that anthropology is a study of embodied material minds and should be primarily about the material social circumstances in which people find themselves and which they negotiate in and through their everyday material practices. It is this that is fundamental to an understanding of how people make sense of themselves and others. Some abstracted notion of mind and infinite creativity does not appear particularly useful to us, hence our stress on a nexus of terms – materiality, embodiment, contestation and emotion –in this study. Kinship, 'village society' or particular social institutions do not reside at its 'core'. Instead we draw on an alternative tradition of material culture studies, as discussed above, together with a holistic and material notion of landscape as its foundation.

Conclusions

Landscapes gather, to use Heidegger's felicitous term (Heidegger 2002: 355ff.). They gather topographies, geologies, plants and animals, persons and their biographies, social and political relationships, material things and monuments, dreams and emotions, discourses and representations and academic disciplines through which they are studied. So landscapes are mutable, holistic in character, ever-changing, always in the process of being and becoming. This book is an act of gathering in which the sum is more than the individual parts. Inevitably we have had to be highly selective and limit the discussions and the detail. Each of the individual chapters might have been developed into a book in itself. The study is an attempt to privilege breadth over depth since any study of landscape requires a holistic approach. The materiality of landscape always outruns us; the real turns into the surreal. We apologize in advance to particular subject specialists who may feel that important contributions have been ignored or overlooked in their own specific field of analysis or discipline of research. The objective here has been to develop a perspective by means of which we can understand landscapes in terms of different performative practices, points of view and modes of embodied engagement. The book is thus a textual attempt to evoke the sheer complexity of the reciprocal manner in which persons engage with landscapes and landscapes engage with them from a variety of personal, moral, social, emotional, ethical and political perspectives.

Part I
The heathland as taskscape

In this section we consider the four main interest groups working on the heathlands today on a regular basis: those who manage and own the landscape; the Royal Marines for whom it is a training ground; environmentalists; and quarrying interests (Aggregate Industries).

2
Managing the Pebblebed heathlands

Introduction

This chapter provides an introduction to the particular landscape discussed in the book and the manner in which it is socially constructed and maintained. It then goes on to consider conflicts that arise as a result of different management strategies and, in particular, those between environmental conservation and historic preservation. It situates these arguments historically and socially, considering what a heathland is supposed to be and for whom it is being maintained, discussing public access and use within the context of environmental policy and politics.

The heathlands (see Figure 1.2) are subject to a web of overlapping managerial control and responsibility of almost bewildering complexity. Most of the land is owned by Clinton Devon Estates (CDE). They are currently managed through a charitable trust, the Pebblebed Heathlands Conservation Trust (PHCT), set up in 2007. The roughly 193 hectares (ha) of Aylesbeare and Harpford Commons, owned by CDE, have been leased to and managed by the Royal Society for the Protection of Birds (RSPB) since 1978, but the RSPB's first involvement on the site came earlier, in 1976. The RSPB has also been working on Venn Ottery Common under management agreements with Aggregate Industries, who own the site. It sometimes gets help with funding from The Aggregate Levy Sustainable Fund for this particular project. The contribution made by the RSPB is of immense importance and its work is managed by a warden and one assistant, helped by two interns who are housed by CDE and by a large number of regular dedicated volunteers. Other areas are owned or leased and managed by the RSPB, the Devon Wildlife Trust (DWT) and the Nutwell Estate. CDE also owns most of the agricultural land in the vicinity of the heathlands and manages these surrounding areas either directly or through tenant farmers. It has a wide variety of other commercial interests in East Devon including management of three business parks, letting of 350 residential properties and forestry enterprises both on the heathlands and beyond.

Parts of the Pebblebed heathlands were first designated as a Site of Special Scientific Interest (SSSI) between 1952 and 1986, as a nationally important example of Atlantic-climate lowland heathland. The East Devon Pebblebed heaths were designated a Special Area of Conservation (SAC) in 1996 under the Habitats Directive of the European Union. The designation covered 1119.94 ha. The primary reason for selection was that the area was considered one of the best heathland areas in the UK because of its combination of north Atlantic wet heaths, European dry heaths, and populations of southern damselfly (*Coenagrion mercuriale*). At the same time the area was also designated a Special Protection Area (SPA) under the Birds Directive because of its rare breeding populations of nightjar (*Caprimulgus europaeus*) and Dartford warbler (*Sylvia undata*). In 2007 the heaths became part of the Higher Level Stewardship Scheme (HLS) administered by Natural England (NE), the governmental conservation agency that runs until 2017.

Since 1963 the heathlands and the surrounding countryside have also been part of the East Devon Area of Outstanding Natural Beauty (AONB), with an agreed landscape characterization document and plan under the local authority control of East Devon District Council (EDDC). The primary responsibility for managing the cultural and historic environment of the area rests with another government agency, English Heritage (EH) and the Historic Environment Department (HE) of Devon County Council (DCC). Beyond these official bodies there are other interested parties concerned with specific management issues, such as the Open Space Society (OSS). The acronyms abound and have multiplied through time, as have the numbers of governmental and non-governmental agencies involved, creating an entangled network of often conflicting policies, people and practices. In the following section we discuss the heathlands from the point of view of key individuals involved in their conservation and management.

Clinton Devon Estates and the heathland

John Varley has been Estates Director for CDE since 2000. Moving from a background in international business and corporate management at British Telecoms, he is responsible for all aspects of management, implementing and developing long-term strategy for the estate's various commercial interests in relation to farming, forestry, property and the heathlands. From a strictly economic perspective the heathlands appear to be worthless land with little or no agricultural or commercial value. Indeed, the need to manage and maintain the heathlands represents a potential drain on resources. Yet, from John's perspective, they are a key aspect of the estate's interests and its entire identity, giving a sense of purpose and direction to the whole. This is because of their geographical and historical significance and emotional value.

Standing on a high point on the heathlands and looking over them – to the agricultural land beyond; as far as the sea to the south and east – virtually all the land one can see belongs to CDE, an entire landscape owned and managed by the estate. From a geographical point of view the heathlands effectively bind it together. They are the spine or backbone of the East Devon Estate (CDE also has substantial land holdings and commercial interests in North Devon and around Beer further to the east). The heathlands symbolically hold together all the other land holdings around their edges:

> There is an integration of forestry, of farming, of agriculture, a whole range of things that interplay with it as well as public access and routes in, routes out, so it represents a core. If you take the core of an apple, if you took that core out and said we don't want this anymore, sold it to somebody or put a fence around it, forget about it, it would be like taking the core from an apple: what have you got left? Some bits around the edge. The whole thing needs the core to hold it together.
>
> (John Varley)

So without the heathlands CDE would effectively become a series of diverse and fragmented agricultural and commercial interests. These might remain of economic value but they would effectively become stripped of emotional value because of their decontextualization from the East Devon landscape. A 'feudal' interest in owning the land as a cohesive, inalienable block, and maintaining ownership of it, is perhaps the key aspect of the entire long-term strategy of CDE. Ownership of such a large area of course provides considerable influence in all aspects of the local economy and the future development of the area in a manner that would not otherwise be possible. This is why the heathlands are not sold, or given away, even to nature conservation bodies such as the

RSPB, who only lease and manage a relatively small area of the heathlands: Aylesbeare and Harpford Commons (for a nominal £5 sum). The rest is still managed directly by the estate itself.

The other, equally important aspect of the heathlands from CDE's perspective is their historical value. Pointing to a series of framed portraits on the wall in the estate office, John mentions that six people have managed the estate (as land agents) since 1865 and that he is the seventh. Beyond that the Clinton family go back to the Domesday Book and have owned this area of East Devon for hundreds of years. This historical connection with the land provides the foundation for the future. In a sense the past is the future, providing a sense of pride, continuity and security. On the website a five-hundred-year history of land ownership is stated, by way of proving that 'rather than being locked in the past it's at the very forefront of visionary business development'.

CDE is run for the private benefit of the Clinton family and other shareholders. The overall return on capital, at about 2–3%, is low compared with what might be gained from selling everything and investing on the stock market – an option that might triple the value of estate assets. However, investment in land and property remains relatively secure in comparison. Because of the geography and the history of the estate it is necessarily not just a commercial enterprise seeking simply to maximize its profits. In a patrician manner the estate prides itself on providing opportunities for local business enterprises, direct employment for over seventy people, homes for local people at affordable rents, and its chain of over 1,000 local suppliers of goods and services. The current emphasis of the estate is on sustainable development responsibly balancing social and economic aspects. For the second time in five years it has been awarded the prestigious Queen's Award for Enterprise in the Sustainable Development category. For the last four years the farming enterprises of the estate have been entirely organic. The estate decides the general policy framework and its tenant farmers have to abide by it. Being organic at present makes good commercial sense since a premium price can be obtained for milk and crops. Some tenant farmers however question whether this will remain a long-term ideological commitment, and the approach of the estate is essentially pragmatic: if it pays to farm organically it will continue to do so. The estate's forestry enterprises are also 'sustainable'. The estate woodlands currently produce circa 10,000 m^3 of Forest Stewardship Council (FSC)-certified timber per year. The produce is sold to manufacturers, sawmills and wood processing industries in local, regional and national markets (Clinton Devon Estates, n.d.). The heathlands and the conservation of the heathland environment form an important part of this strategy of sustainable rural development. According to John this is very important in creating the desired image and impression of the estate in the eyes of the general public. Managing the heaths for nature conservation is therefore important in terms of the wider general image management of the estate's commercial enterprises and interests:

> In terms of what our ethos is to the environment, to have something as precious as that, to be in your stewardship, other companies would bite your arm off [to have that], they do, they fund water schemes in Ethiopia to help poor people, they might find something else or a nature reserve. We've got one and it's here.
>
> (John Varley)

In the past this potential of the heaths to be incorporated into a wider business model balancing economic, environmental and social policy was insufficiently realized. Current CDE policy is therefore to put the heaths to work, as it were, in terms of the broader commercial interests of the estate. In a nutshell, CDE aims to be good for the Clinton family as a profitable business, because it is good for the local environment and good for the local population. It aims to be a shining beacon in an otherwise somewhat tainted national and international corporate world of twenty-first-century feral capitalism.

This has not always been the case. From the 1920s onwards the afforestation of large tracts of the heathland was undertaken in order to make commercial gain, at least in the long term, from what was otherwise useless and infertile land. In 1971 the estate made a planning application to create two championship golf courses on Woodbury and Colaton Raleigh Commons. The area of heathland to be transformed was huge, amounting to 505 acres (204 ha). These would occupy the highest and most prominent areas of the heathland to the north and east of Woodbury Castle, and would come complete with a substantial club-house and other facilities (see Wilson 2004 for an account). This caused widespread public protest and demonstrations, including a campaign to save the heathlands from destruction, which was to last for three years until the planning application was finally refused by the Secretary of State for the Environment in 1974 (having been approved by Devon County Council contrary to its own development plans).

The establishment of the Pebblebed Heaths Conservation Trust (PHCT) in 2006 as an independent charity was a CDE initiative to underscore its *lack* of commercial interest in the heathlands. At the same time the estate entered into the HLS scheme for its management. CDE had long received small grants for conservation work from various sources, including the Countryside Stewardship Scheme and lottery grant money under a Heathland Heritage scheme. The new HLS scheme is a relatively long-term one operating until 2017 and providing a subsidy of £200/hectare. Given the substantial land holdings this amounts to a current annual income of around £224,000 for conservation work. From the point of view of the estate this income is not sustainable in perpetuity. Once the scheme runs out the heathlands may once more become a drain on estate finances, and their future in the absence of a long-term scheme remains effectively insecure. Given the long-term commitment of the estate to the heathlands, all the schemes available remain unreliable because they operate only on a short-term basis and because political priorities and funding opportunities change.

Despite the apparent lack of economic potential the heathlands have in fact always provided and continue to provide the estate with substantial income streams that are not used for environmental conservation. These include income from leasing land to the Royal Marines for a grenade range, and a licence for the Marines to train in the area (c. £60,000/annum); very substantial but undisclosed income from open cast quarrying operations at Black Hill that have only recently finished; income from forestry interests and from rent of engineering works in the quarry area; and small sums from the model aircraft flying club, Dalditch stables and others for use of the land in various ways. The estate is currently investigating the possibility of establishing car-parking fees in order to contribute to PHCT funding.

Visions of the heathland and their management

In the next sections we look at some of the key people and organizations actively involved in the management of the landscape.

> It's not a manicured, managed habitat or landscape, but it's not bleak and open and rugged and wild but it's, you know, a little bit in between, so you get the best of both worlds really, you sort of get the fact that you've got the accessibility and things and yet at the same time if you want to, get away from it all, you can retreat into some quiet bits.
>
> (Toby Taylor, RSPB warden)

The RSPB has had a major input into the management of the heathlands through its work on the Aylesbeare and Harpford reserve and through the restoration of the Black Hill quarry site, along with work for CDE elsewhere across the heathlands that it carried out under the old government-funded countryside stewardship scheme or that was funded through the National Lottery. The reserve area was first 'gardened' by RSPB staff during the summer months from the mid-1970s. A full-time post of Annual Warden was established in 1985. Three different wardens preceded Pete Gotham, who took over in 1988 and was warden for eighteen years. Originally a quantity surveyor, Pete's passion for conservation led him to a major career change, first managing RSPB reserves in the east of England before moving to Devon, where he initially lived on site in a caravan throughout the year. It was Pete Gotham who was responsible for building up the resources of the RSPB reserve from virtually nothing to comprising a full suite of tools, machinery and vehicles and a team of volunteers. He was also responsible for the initial landscaping of the reserve, establishing ponds, cleared areas and bare ground, and introducing new techniques for its management: cattle grazing, bracken bruising with machinery and topsoil scraping (discussed below).

The present warden, Toby Taylor, has continued this work of improvement and enhancement. His family moved to Devon when he was just a few weeks old and he was brought up in Exmouth. He spent most of his time at weekends out on the heath, first on push bikes and later on motorbikes, exploring it and enjoying the landscape. Following a course at the local Bicton College of Agriculture he decided to specialize in conservation work and started volunteering with the RSPB in 1992. He was eventually offered short-term contracts before obtaining the post of warden on Pete's retirement in 2004. Like Pete he has a passion for nature conservation and for this particular heathland landscape, which he regards as distinctive and unique on account of its location and geology. His vocation is to make it as perfect as possible from the point of view of wildlife conservation, and to maintain and further create an intricate landscape mosaic. He now organizes a team of two full-time members of staff, people on short-term contracts and up to fifty volunteers to do the work (see below). He creates a five-year independent management plan for the RSPB, but more important is the HLS agreement with Natural England (NE) since this actually pays for the work and because it is the targets set out in this agreement that will be assessed. The RSPB management targets are if anything higher or more stringent than those set out by NE. Irrespective of the meeting of artificial targets Toby takes great pride in his work: 'I'm very, very lucky to have an influence and be able to work in an area I've grown up on and had a great interest in ... [and have] an appreciation of the whole.'

From car parks to conservation site

Bungy Williams (recently retired) was Commons Warden for CDE for over twenty years. He knows the heathlands better than anyone else and his own biography is very much bound up with them. Born in Hamworthy, Dorset, his first encounter with this landscape was in 1965 when he joined the Royal Marines (RM) as a sixteen-year-old recruit and did his basic training on the heathlands. Apart from the pain of the constant runs and having to negotiate the trials of the endurance course (see Chapter 3) he remembers little in detail about those raw early experiences of the heathlands. After twenty years of active service he returned to the RM base at Lympstone as a Sergeant Major and physical fitness trainer. He eventually gave up a well-paid and secure job to accept the post of Commons Warden, at half his RM salary. Before his appointment in 1990 there was nobody who had a full-time responsibility for looking after the heathlands, and conservation of the habitat was hardly on the agenda. Bungy was felt to be an ideal candidate for the job because as an ex-Royal Marine he knew the area particularly well. Besides this, because the Royal Marines use the heathlands extensively in their training, it was expected that Bungy would know what the Marines were doing, and what they shouldn't be doing, and how to interface with them. When Bungy took up the post there was little in the way of a job specification, no management plan, and only a few hundred pounds to pay for limited scrub clearance. The work primarily involved checking car parks, rubbish clearance, looking after signage, access and gates:

> My job was to look after the car parks. The guy I actually took over from was part time. He had his own vehicle so the Estate never even supplied a vehicle and he used to work in Lord Clinton's garden. He used to own one of the fish and chip shops in Exmouth, and he was just retired and he liked it up here [on the heathlands], and started clearing the rubbish for two to three days a week, things like that. I think there had been a couple of pre-wardens ... But I thought I can't be doing this for much longer. There's no brain in this at all!
>
> (Bungy Williams)

While serving in the RM Bungy developed an interest in the natural world, sparked by adventure training with recruits in South Wales whom he taught to identify different plants and so forth, something that was very definitely not part of the agenda during his own training on the Pebblebed heathlands back in the mid-1960s. Talking with Pete Gotham, the RSPB warden, encouraged him to learn much more and actively develop his own wildlife conservation agenda for the much larger area of the heathlands for which he was now responsible, which is precisely what he has been doing ever since. Although parts of the area had been designated a conservation site since 1952 when Bungy started work, he did not even know who the conservation officer was: 'I never met him. I met the second one twice in two or three years. It's only just recently during the last five, six, seven years when they have actually been getting more involved with it.' Bungy received the MBE for Services to Nature Conservation and the Environment in 2004.

Two very different management styles take place on the heathlands, one extensive, using machinery, the other intensive and using predominantly hand tools and human labour. On the Aylesbeare and Harpford RSPB reserve a lot of the work is carried out by a team of up to fifty volunteers (see Chapter 4). On the rest of the heathlands, managed by the PHCT, it involves the use of heavy machinery, such as a tractor with front-mounted and rear-mounted tungsten bladed mowers to manage vegetation and scrub. We asked the RSPB warden and that of the PHCT to take us across the land that they manage and show us in practice what they do. This involved a 2 km walk across Aylesbeare Common with Toby Taylor, as opposed to a 12 km Land Rover drive with Bungy Williams with numerous stops along the way to look out at the landscape. These modes of encountering the landscape in themselves well exemplify the two different management styles.

The walk with Toby took in areas of wet and dry heathland, artificially created ponds, and areas that were once woodland, now felled and maintained by livestock grazing.

Managing heathlands with cattle is regarded as a much more 'natural' approach than using machinery. The cows' grazing creates random patterns of vegetation removal that are difficult to achieve with machinery. However the cattle 'naturally' prefer to graze in the mires and tend to avoid dry heathland unless fenced in. Cattle can't be fed during the winter on the SSSI and, as non-native grass seeds cannot be introduced on an SSSI, more and more parcels of land around the heathland are required to maintain them during the winter. This means that in order to manage the heathland in the style required, surrounding areas are needed as

Figure 2.1 Wet heath, Aylesbeare Common

Figure 2.2 Dry heath, Aylesbeare Common

Figure 2.3 Cherry trees

well with artificial or cultivated grassland. This in effect accentuates the 'island' character of the heathland areas and the borders between it and non-heathland areas. Unwanted vegetation such as birch may either be clear felled or killed with herbicides and left standing to be colonized by invertebrates. Much of what is done is not planned in advance by looking at a map but improvised and decided on the spot, and this often involves aesthetic judgements about what looks best in a particular place.

It is often the people working on the ground, particularly experienced volunteers who have been working on the heathland for anything up to twenty years, who make particular decisions about what to do. For example, cherry trees grow on one area of Aylesbeare Common, planted in the 1980s when this area was heavily covered with birch scrub, a decision taken by a local volunteer.

These trees still remain despite the fact that they are hardly characteristic of heathland, while the birch has long since been removed. In this respect the management structure is personal and relatively non-hierarchical in character. So there are lots of different individual inputs into the labour-intensive management of the RSPB reserves as opposed to the rest of the heathlands, where the same few people make the decisions as to what to do. In this sense the PHCT-managed areas are much more reflective of the personalities of just a few individuals and their interests and values. In the 1990s water for firefighting was a major concern on the heathlands, with numerous ponds being created then, but the unintended consequence was an increase in biodiversity.

This heathland is a landscape layered by names discussed in relation to different user groups and interests throughout the book. One of the primary ways of relating to and remembering places is to name them (see e.g. Basso 1984 and 1996 for exemplary studies of western Apache place names; Gaffin 1993 and 1996 for the Faroe Islands, Kelly and Francis 1994 for the Navajo, Weiner 1991 on Papua New Guinea, Laviolette 2011a: 93–115 for Cornwall). Basso shows how names enforce moral narratives. The naming of places is fundamental in the manner in which people can be 'stalked by stories'. They cement place and event in a powerful way through the manner in which the name itself (often a sentence) acts as a precise description of place. Kelly and Francis (1994) similarly show how for the Navajo places are part and parcel of stories and narratives linking the living and the dead, past and present, mortals and the immortal and spiritual powers that make the land sacred. Weiner (1991) emphasizes the manner in which, for the Foi, names create place out of an empty void of space or environment, and humanize place in the most profound way as an integral part of personal biographies and movement along paths of movement connecting one place to another: place as journey. Gaffin (1993, 1994) links the use of place names to an ethos of environmental conservation and argues that they are intrinsic to its protection and the individual and social identities of the Faroese. Platial and social order are intimately linked. The manner in which people 'place' themselves is central to an understanding of local ecology that involves a sense of and a feeling for place. Laviolette (2011) discusses the manner in which Cornish place names have multiple references to a Celtic past and sense of distinctiveness, relations between insiders and outsiders, a sense of directionality and topographic features of the landscape

On the Aylesbeare and Harpford reserve all the ponds have individual names given to them by the environmental volunteers, e.g. the gas pond, built along the line of a gas pipeline, Tom's Pond, Potters Pit, etc., but these are known only to RSPB members.

Figure 2.5 shows the thirty-nine names used by RSPB staff and volunteers superimposed over the topographic map with its standard names for a small area. A wealth of personal knowledge, events and associations is revealed. Places are named after important natural topographic features, such as ravines and the numerous ponds and old quarry sites that have been artificially created, wet and dry areas of the heath and

Figure 2.4 Gas Pond

their characteristics ('wet finger' – a linear boggy area), woods, plantations and types of trees. Other names refer to built structures, houses and gardens. The places are also intimately related to persons: the former warden's caravan is marked; one can find Nicky's bit, Tom's pond, Basil's pond, etc., and there are references to orientation points such as the 'lone pine', a dead pine-tree trunk standing on a prehistoric cairn that used to be a very prominent orientation point for anyone walking on this and other areas of the heath. It was visible even out to sea. Known also to some as the witch's tree, it is now fallen in a storm.

    The printed topographic map with its names is only a skeletal, official, abstracted construction. These place names, unlike most of the map names, reveal the manner in which this part of the heathland becomes humanized and historicized by those who work in it, in relation to their activities and how they perceive the landscape itself, what is important for them in relation to the ecology. As mnemonic devices they are part and parcel of place-making activities, of the construction of landscape and the establishment of their own personal connections to the land and those they have worked with.

    The Land Rover journey with Bungy over the PHCT-managed area of the heath was far more varied in terms of the landscapes and places

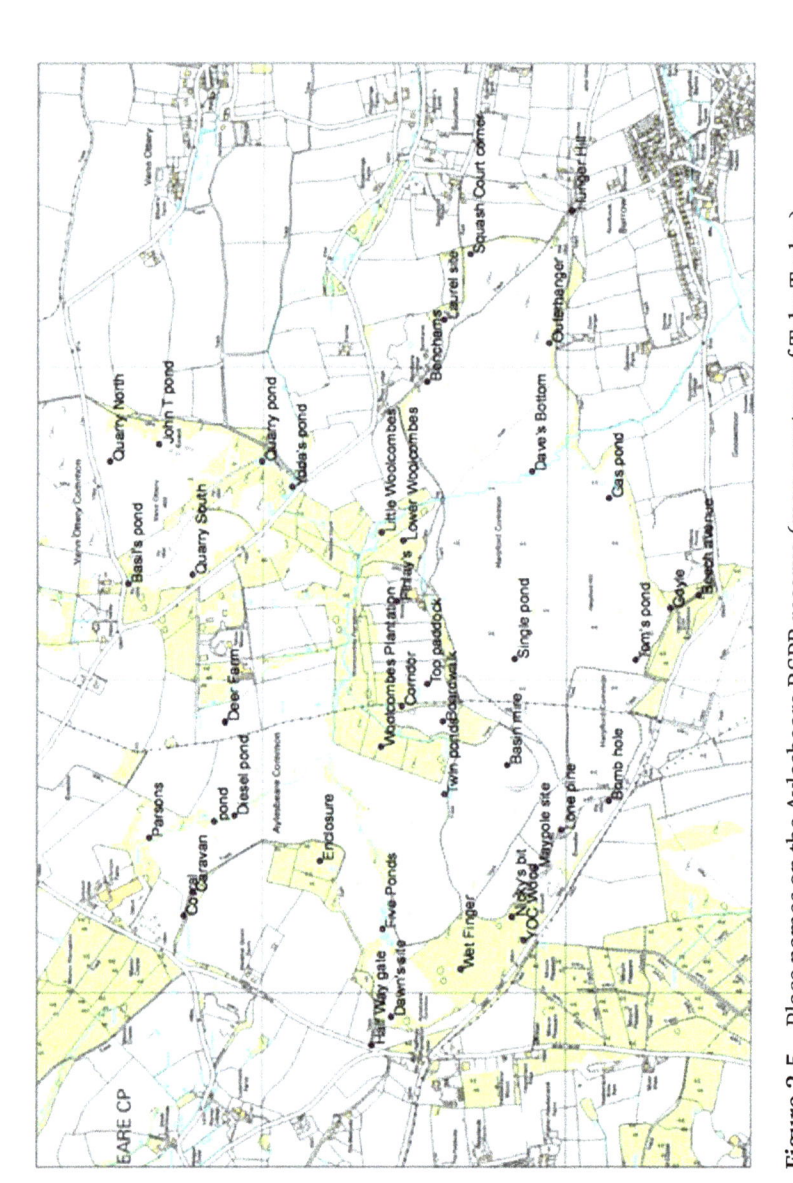

Figure 2.5 Place names on the Aylesbeare RSPB reserve (map courtesy of Toby Taylor)

Figure 2.6 'Savannah' heathland, Woodbury Common

covered. We were shown old car parking areas, now closed; a Second World War dumping site, with debris still exposed at the surface; wetland areas with Red Devon cattle fenced in and grazing; old quarry pits; grassland areas extensively burnt a few years previously; old tree ring enclosures; a military bunker from the Second World War; an underground water reservoir used by the fire brigade with a Victorian pipe still in use to feed the lake in Bicton Gardens; a 'savannah area' with pine trees dotting the landscape in a picturesque manner (not strictly conforming to NE pine-tree density rules), but much appreciated by the general public, (see Chapter 6); swaled areas with 'snake' bends; so-called degenerate and pioneer heath areas; patches disturbed by digging by the Royal Marines; remains of the old Second World War army camp at Dalditch, with brick and concrete hut foundations; the Iron Age hill fort of Woodbury Castle and so on – a huge area managed by just two people.

Bungy has been responsible for coordinating pond-digging and the construction of walkways and steps (the latter two with the help of young offenders on probation) in various areas across the PHCT-managed areas, but none have place names given by him or his assistant. This is because Bungy continues to use the Royal Marine names instead (see Chapter 3).

It is no exaggeration to say that the heathlands that we see today are very much bound up with the biographies and perspectives of a few people and can be regarded, in part, as their personal and collective creation and vision of the heathland. This is explored further below.

Figure 2.7 Toby Taylor's map

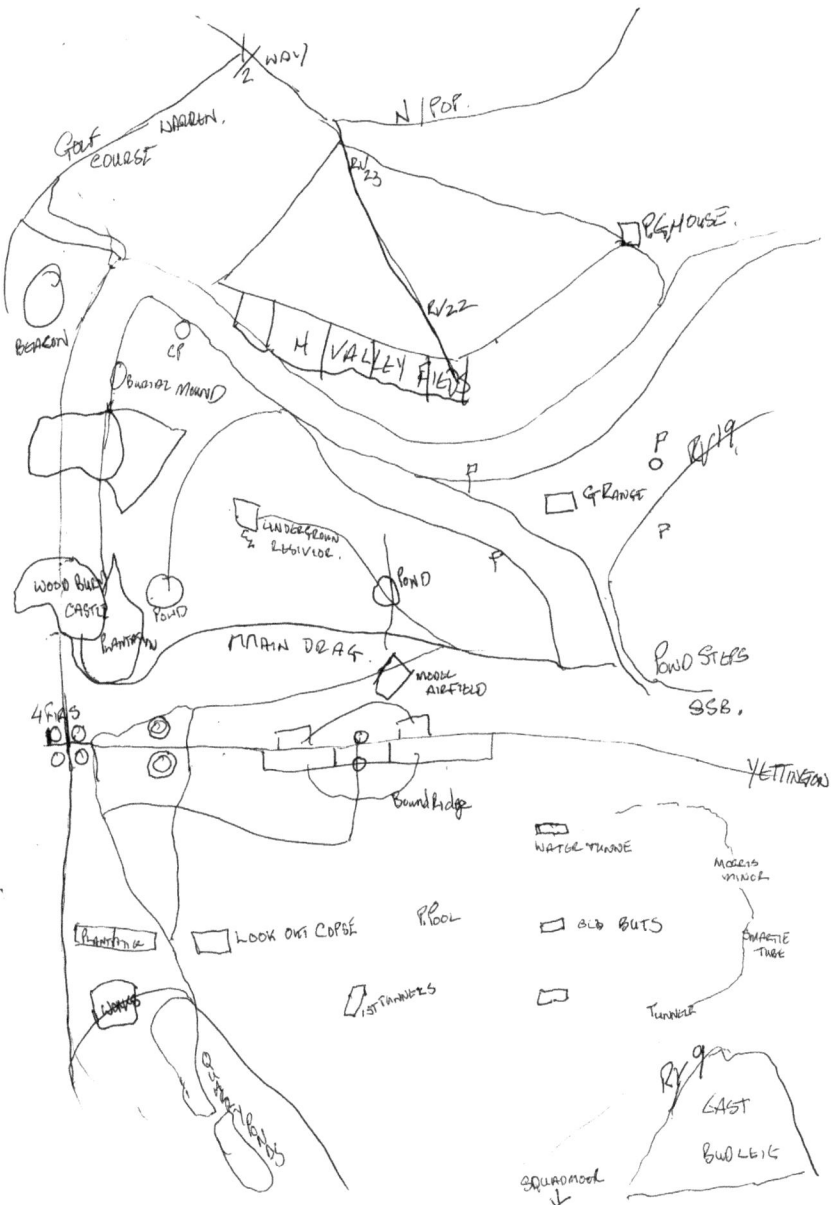

Figure 2.8 Bungy Williams' map

The maps produced by the two heathland conservation managers differ quite significantly. Toby's map (RSPB) emphasizes administrative divisions and properties, and shows the roads on either side and across the heathland from west to east with about sixty features marked. Different commons are named, as are villages, hamlets and individual properties such as farms.

Woodbury Castle, the landscaping mounds at Four Firs, Bicton College, the quarry, reservoir, the model aircraft flying field (hereafter MAFF), and the Bronze Age barrows at the summit of Aylesbeare Common are marked. Some wooded areas are shown. The RSPB reserve is shown but otherwise the only wildlife mentioned are bats on East Budleigh Common.

Bungy's map interestingly shows the heathland from the perspective of both a RM and a conservation manager.

Parts of the RM endurance course are named and marked and various numbered RM rendezvous points. The grenade range is shown, together with its surrounding marker flagpoles and Lookout Copse, another RM place name. Plantation areas are marked, together with ponds, quarry ponds, walkways and main tracks. The landscaping mounds are shown along the Woodbury to Yettington Road, with one prehistoric barrow.

Perspectives of conservation professionals

Natural England (NE), formerly English Nature, has the responsibility for managing the current HLS scheme and as such effectively decides the general principles and practices of heathland management, while the actual work is carried out by the PHCT and the RSPB. It has its local regional office in Exeter and the HLS scheme for the heathlands is only one of many that are administered from there. The personnel involved from NE are often given different areas to look after at different times and have multiple responsibilities with regard to different conservation areas. This means that they often cannot dedicate their time to a single scheme, conservation area or nature reserve and so it is necessarily the case that the person or persons involved have a limited first-hand knowledge of these landscapes. In practice what this means is that they need to be shown and guided through these landscapes by local people involved in conservation work who do have a deep and intimate knowledge of them. So, overall conservation priorities are decided upon in an abstract and theoretical way guided by map-based knowledge, assessment reports and fieldwork carried out by others: this is, largely, management from a distance and from behind a desk. To set priorities for managing and conserving a landscape while lacking in-depth personal knowledge of it might appear rather odd but this is the way in which NE and environmental consultancy in general operates. Having sufficient theoretical and ecological knowledge is all that is apparently required to manage an area. Knowledge of the actual material landscape is a secondary consideration and a low-level priority. Since all NE staff concerned with conservation management have the relevant degrees and diplomas they are treated as interchangeable with regard to the casework side of things and assessment surveys.

A major report commissioned by the PHCT and published in 2009, examining options for the future management of the Pebblebed heaths (see below), is a good example of this (Underhill-Day 2009). It contains no new research information but very usefully summarizes and brings together existing knowledge about the Pebblebed heaths from a conservation point of view. It also discusses in detail general considerations with regard to heathland management based largely on research undertaken on heathland areas elsewhere in England and parts of lowland Europe. What actually makes the East Devon heathlands special and important, it might be argued, is the *specificity* of the locality: its unique pebble geology, the local topography and climate, its situation next to the sea, its archaeology and historical development. All this is virtually ignored in the seventy-seven-page report. For example, twenty-three lines discuss the archaeology of the area with no mention of the need for managing and conserving historic or archaeological features or developing a management plan for them. The geology, soils and topography are all dealt with in another twenty-three lines, resulting in a rather limited discussion highlighting nature conservation issues. The rest of the report, with an additional forty-two pages of appendices and maps, concerns itself with descriptive summaries of the plant communities and wildlife of the heathlands and management conservation issues. As a consequence of this the 'heath' in the report becomes an abstracted category rather than a term with local meaning, value and relevance. It might be a heath anywhere. The only real local specificity concerns estimated numbers of rare bird species, such as the hobby, Dartford warbler, and nightjar, and information about various dry and wet (bog and mire) heath plant communities. Even the spatial distribution of the various heathland categories (favourable, unfavourable declining, unfavourable recovering, gorse dominant, heather dominant, etc.) are not plotted on maps of the heathland in the report. These maps show only the SSSI boundaries and the commons boundaries, and recorded nightjar and Dartford warbler territories. As a consequence, their spatial positions and relationships that might be regarded as important in terms of an informed local discussion of heathland management strategies remain unknown to the reader.

But what is largely forgotten here is that such management ideally requires establishing long-term social relations with those who carry out the actual work. It is local people who do the work on the ground and who may feel a sense of alienation with regard to top-down management priorities set by an aloof government agency. Furthermore, institutional memories of particular problems and perspectives in the practice of managing a particular conservation area are often short, not surviving the appointment of a new official, creating a sense of having to start all over again. The current Conservative government-imposed austerity regime that is being imposed, with NE staff reductions, seems likely only to make the problem worse.

The management scheme for the heathlands is thus derived from a general model of what heathlands in general, and lowland heathlands in particular, should look like, with some allowance for local differences, such as species that are not found elsewhere. The mantra is to improve them until they reach the 'ideal' state as specified by the model. So this does not just require maintaining the heathlands as they are and conserving what is already there, but enhancement, the overall aim being to increase current levels of biodiversity. What is required is the creation of a mosaic of different vegetation types across the area. In this way, it is hoped, niches will be created to allow different species of birds, insects and other invertebrates to inhabit the area and thrive there. NE sets targets for heathland management and carries out condition assessments of the heathland, using generic assessment forms which are recorded in the field on a rolling basis from surveys of one- or two-metre quadrants in eighteen different heathland assessment areas. In this way it aims to measure the extent to which management practices achieve the targets that have been set. These must be quantifiable and measurable targets, but much that is valuable about the heathlands to people simply cannot be quantified in such a manner. A condition assessment of the heathland carried out by NE between 2002 and 2008, with reference to 18 SSSI units, provided the results given in Table 2.1.

The majority of the heathland is thus classified as being 'unfavourable' but recovering toward a 'favourable' status, while six per cent is in 'decline'. The main reason for designation of much of the heath as being 'unfavourable' is lack of a sufficient level of management, so that much of the heath has an undifferentiated age structure in terms of vegetation growth (e.g. pioneer heather, building heather, mature heather, degenerate heather) and because of encroachment by scrub and bracken.

Table 2.1 Proportions of different heathland types

Condition	Area	Percentage of total heath
Favourable	0.32ha	0.03%
Unfavourable – recovering	1065.62ha	93.89%
Unfavourable – declining	69.05ha	6.08%
Total	1134.99ha	100%

Source: Underhill-Day 2009: 18

NE, like all other UK government agencies, has been subject to an audit culture (see Strathern 2000) since the 1990s. This is a managerial rationality requiring transparency, accountability and value for money in decision making. The only knowledge considered worthwhile within such a framework is that which can be measured and quantified in relation to performance targets that shape the behaviour of both institutions and those working in them. In relation to the landscape we can do no better than quote Carter's poetic words:

> Thought that identifies knowledge (and power) with the achievement of a rational *terra firma* – with cut-and-dried definitions and their corollary, instrumental scenarios for planning the future – [It] is an act of intellectual aggression towards the environment of thinking (recollection, imagination and invention) in general. It treats the humid, life-giving zones of creativity as swamps and morasses ... The desiccation of the planet may be partly due to anthropogenic environmental practices, but it is legitimated by a dry thinking that assumes the only good ground is flat and dry.
> (Carter 2010: 11–12)

The audit procedures that NE undertakes for the heathlands are environmental condition assessments. They are essentially form-filling exercises, enabling the measurement of targets according to pre-defined and essentially arbitrary categories. One NE official put it this way:

> You can't have a 'favourable declining' [category] although its often been mooted there should be a 'favourable declining!', but you can have a 'favourable recovering'. So if you put down something that is 'unfavourable recovering', effectively what that is saying is that – although it is actually failing the condition assessment – the management that is in place is probably the best that you can do and it's, you know, in the long run, be it ten years or a

hundred years, that habitat is gradually moving to what it should be moving towards.

(Tom, NE advisor)

But the problem posed by the audit culture is that measurement of target attainment is always and necessarily short term, as is the government funding for the achievement of generic targets for habitat types such as heathlands formulated by a body of experts who make up the Joint Nature and Conservation Council (JNCC). Overall conservation objectives set by NE are to maintain and improve dry heath and lowland wet heath (bog or mire areas), and habitats for populations of rare bird species (Dartford warblers and nightjars) and the southern damselfly. This involves:

- ongoing control and management of woodland, scrub, gorse and the creation of bare ground;
- management of plant communities to achieve a mosaic of different ages and structures across the vegetation of the heaths for the benefit of the associated flora and fauna;
- control of spreading grasses at the expense of the dry and wet heath and mire communities;
- providing a range of facilities to improve people's enjoyment and appreciation of the heaths and managing visitor pressures to minimize the impact on wildlife and the environment.

Translated into human, or anthropological terms, management of the heathlands – as is also the case for conservation management elsewhere in the UK and globally – might be described as a version of 'ethnic cleansing'. Unwanted species are removed, in this case primarily bracken and trees and scrub of all kinds, in favour of others. Even in the case of favoured species the proportion of them and their age has ideally to conform to prescribed guidelines. For example, the tall European gorse should not exceed 25% of coverage in assessment units.

The archaeological perspective

From an archaeologist's perspective, the heathland is a precious resource because much of it has never been ploughed, and therefore archaeological remains have been left undisturbed, contrasting with the vast majority of lowland UK landscapes. The introduction of the HLS scheme in 2007 meant that specific provision for the preservation of historic sites in heathland management had to be taken account of in a way that had not happened before. Specifically funding was made available for the restoration of the major and most important archaeological site on the heathlands, the Iron Age hill fort of Woodbury Castle. The castle's ramparts had suffered considerable erosion scars in places through visitor pressure, including both people running up and down the ramparts and the activities of some mountain bikers. Since Woodbury Castle is a scheduled or protected monument the work was administered through English Heritage (EH) and involved scrub clearance, infilling of damaged areas and the provision of floating stairs allowing access to the interior on established pathways.

The most recent survey undertaken by Exeter Archaeology (2003) identified 341 surface archaeological features across and in the immediate vicinity of the heathlands. These range in date from Bronze Age barrows or cairns to remnants from Second World War military training. Only a few of these sites are scheduled, or in other words have any legal protection. All have suffered neglect through the lack of scrub management in the past, though scrub has now been cleared from a few of the major scheduled monuments. Given the character of the heathland vegetation many sites remain to be discovered.

In comparison with the funding and resources made available for environmental conservation, historic conservation has been very much a Cinderella concern. For NE the heathland landscape as a whole is significant, with its myriad and changing vegetation patterns and wildlife communities. The historic resource is, by contrast, considered in a very different way: essentially a dots-on-maps approach. Field surveys reveal sites that are recorded in the Historic Environment records (formerly the Sites and Monuments record) and registered on maps kept locally by DCC and nationally by EH. Those deemed the most significant and of national archaeological importance have been scheduled and are protected from destruction by law. A line gets drawn around them on the map and anything beyond that line – areas that may be only a few metres distant from the visible extent of a monument such as a round barrow – remains unprotected. This approach effectively disregards the landscape settings of archaeological and historic sites and monuments and their relationships to each other across the landscape. It disconnects them from their landscapes. What makes the heathlands so significant today from the point of view of both environmental and historic conservation is that most of the land has not been cultivated. The invisible archaeological resource (that which lies beneath the surface), as well as the visible, is still preserved.

What time is this heath?

Ultimately what the heathlands look like and whether they appeal to visitors, who might then regard their conservation as important, depend on a vision of heathland and what it should be like in which targets and an audit culture often seem less than helpful. Conservation management of the heathlands promoted by a government agency, NE, has an inbuilt central irony in that this is very much a cultural or human-created landscape, and in this sense there is nothing whatsoever that is 'natural' about it. An unnaturally natural landscape is being preserved and, like all landscapes, it has changed over thousands of years primarily as a result of human activity. Furthermore, nobody really knows in any geographical or environmental detail what various parts of this landscape were actually like 100, 200 or 1,000 years ago, because of a lack of detailed historical records or archaeological evidence: where the gorse grew tall, the heather was thick, where bracken and grass dominated or scrub was prevalent, what species of birds were present, etc. Furthermore, conservationists are well aware that climate change is likely to alter which species inhabit the heathland in the future. Change rather than stasis is the norm, but the latter perspective essentially governs the creation of targets: 'that's what governments want you to do and that's how people work, you have to have targets, so you have something to measure' (Tom, NE advisor). From a perspective in which historical and environmental change are acknowledged, targets specifying percentages of various vegetation types seem quite bizarre. Preserving heathlands is always a historical issue involving the question as to what period in time the conservation and preservation of them is supposed to relate to.

Time is thus a largely hidden or fourth dimension relating to heathland management. In relation to the specific objectives of heathland conservation John Varley is rightly somewhat sceptical about precisely what is being managed and conserved and why:

> My question to them [conservation bodies such as NE] has always been: what do you want? Heathland? But is that 1850 heathland, 1896, would you like it as it was in the 1700s? Do it just after World War II, in 1950, a very different landscape when the tanks and vehicles have just been taken off the ground, we have aerial photographs, a very different landscape. We can deliver 1970, 1980 where it is just a mess. I said tell me what you want and they go ah! They can't answer the question!
>
> (John Varley)

Figure 2.9 Vegetation scars caused by pit digging by the Royal Marines, Colaton Raleigh Common

Bungy Williams is very much aware of the specificity of the landscape in which he has worked for over twenty years. Although he now receives regular advice from Natural England about what to do and how to manage the heathlands under the HLS scheme, he is the person who actually implements the plans on the ground and decides where to work, what to do, and how. For Bungy these may be lowland heathlands, and in this sense superficially similar to other lowland heathland areas in the UK, but at the same time they are the East Devon Pebblebed heathlands, a unique landscape in a unique geographical location, with a unique history of use and a unique connection to his own personal biography. For him this ideally requires a rather different, locally sensitive and attuned conservation agenda distinct from that developed for heathlands in general and then applied in a top–down fashion to this area of East Devon. As we have seen, much of this heathland area does not look the way a heathland should look according to Natural England. A lot of it is classified as 'unfavourable' and some of it as 'declining', but this is only from a particular institutionally defined and abstract point of view. Some areas of the Pebblebed heathlands will never achieve the NE targets because effectively they will always be the wrong kind of heathland. This is partly for historical reasons. In the Dalditch area, in the southern part of the heathlands, the building of an extensive Second World War camp housing up to 5,000 men involved the importation of non-local materials, such as lime, irrevocably altering the character of the vegetation even though for the most part only foundations remain. Consequently individual species and plants not supposed to be found on heathlands grow here. Another case in point is that in other areas of the heathlands, to the east of Woodbury Castle, there are extensive areas that have been affected by the digging of trenches by the Royal Marines, which has altered the character of the heathland vegetation. Typical vegetation never grows well in these areas and there are numerous circular or oval patches of 'unfavourable' heathland as a result.

Elsewhere proportions of heather and gorse areas are 'wrong'. Bungy comments:

> I think their [Natural England's] knowledge is about heathlands, not about a heathland, and people used to call this grasslands … My worry is we will become a mosaic of what they want and lose what we've got. I can point to things like the silver studded blue butterfly, Dartford warbler. The Dartford warbler needs that gorse out there and we've got a good population of them. It is because of that gorse out there that NE want to get rid of, well not get rid of it, but they don't want as much, because they think if we get rid of some of this we might get something on the ground. Great if you get it but what happens if you don't get it? You've then got rid of the Dartford warbler which is a listed species purely because you want this complete mosaic.
>
> (Bungy Williams)

Bungy conforms to management plans for reducing heavy gorse cover in some areas but does this in his own particular way: 'You've seen where I have been chopping, just behind us here [we are talking in the Estuary car park near to Woodbury Castle], but I don't do it as intensely as they [NE] might like.'

The required mosaic of different heathland habitat types is delivered and organized on the ground by the Commons warden for the bulk of the heathlands and by the RSPB warden, Toby Taylor, for Aylesbeare and Harpford Commons. It is these two men and the legacy of the former RSPB warden Pete Gotham that actually influence the physical character of the heathland that people experience today despite the NE directives. The main management techniques used for species removal and manipulation are topsoil scraping, mowing, swaling or burning, herbicide spraying or

Figure 2.10 Heathland management mosaic: Aylesbeare and Harpford Commons

cutting and bruising of bracken, removal of encroaching scrub and trees and limited grazing by animals (cattle and some ponies in selected mire areas). Some of these are further discussed below. Other maintenance activities involve the provision of access and information for the public, and dealing with problems that arise from such access: maintaining car parking areas, gates, stiles, footpaths, signage and information boards, firebreaks, bridges over streams and walkways in boggy areas, removal of litter and fly-tipped material. Currently management of the heathlands takes place in two main seasonal cycles. During the autumn and winter months (September to March) scrub and tree clearance, cutting, mowing and swaling take place in selected areas, together with some topsoil scraping. During the spring and the summer firebreaks are cut and much time is spent on cattle management in restricted wetland areas, maintaining fencing and so on. The management work is described by all involved as very much an uphill struggle: no sooner has scrub or bracken been removed from one area than it grows up somewhere else.

The outcome of all this management is that, walking over the heathlands, a visitor can thus experience small, artificially created ponds in bog and mire areas along stream valleys with a wealth of aquatic plants and insect life. Ponds like these do not occur naturally on the heaths.

Figure 2.11A Harpford Common: swaled areas (light circular areas in background)

Figure 2.11B Heathland management on Woodbury Common: mature heath (left), newly cut heath (right)

Figure 2.11C 'Snake bend' firebreak with path, Colaton Raleigh Common

Figure 2.11D Artificial pond, Colaton Raleigh Common

Some are even provided with small islands for nesting birds. In dry heath areas there are bare patches where all the soil and vegetation has been scraped off, both in summit areas and on hill slopes. The size and character of these scraped areas is related to topography, tree cover and other factors such as where to dispose of the scraped material. In other areas of the heathland there are circular, lunate or irregularly shaped swaled and mown patches: 'we do a circle, a moon shape or we do any shape, you know, just so it looks completely haphazard' (Bungy Williams). Curving firebreaks sweep across some areas, mown through the gorse and heather. Tall gorse patches are retained in some places with clear-cut or mown areas in between. Individual pine trees are allowed to grow in some places, breaking up the heathland vista, while in others small circular or irregular groups of trees are left where scrub and woodland clearance has taken place. Wooden walkways are provided across some mire areas and bridges across some of the deeper and faster-flowing streams.

Straight or geometric edges are generally avoided except where they are already in place along plantation boundaries. The conservation heathland mosaic created in this fashion is meant to be picturesque and pleasing, giving the landscape a 'natural' and 'wild' appearance. Heathland management, whether intensive or extensive, thus aims to disguise itself in its informality, through the creation of this improved 'timeless' landscape. So successful is this that most visitors are not aware that everything they see and experience has been created. Sometimes the work that is actually to be done is planned and marked on a map in advance. Often it is improvised during the process of actual management work. The inclusion of 'traditional' breeds of livestock, red Devon and black Galloway cattle together with a few hardy Exmoor ponies completes the picture perfectly: we are in a version of the past created in the present. The 'past' thus created is essentially idealized and romantic, forming a landscape that never was. The heathland in effect becomes a vast landscape garden, nature imagined and improved. In this manner abstract management targets for heathland and vegetation types dreamed up by officials in an office get translated into an artful and artistic practice on the ground into a conservation mosaic that is enticing for leisured walking and wildlife appreciation with little secrets, a pond here, a small copse there, a graceful pine over there, to be discovered along the way.

This is a very different kind of heathland from the often harsh realities of the intensive economic and agricultural exploitation to which it was historically subject (Vancouver 1969). The contemporary heathland that is being created is about as faithful to that past as a Disneyland fantasy, but it actually has no pretensions to create a version of the past in the present. The heathland we see is a contemporary creation, a twenty-first-century vision of heathland, and is none the worse for that. The romantic management of heathland tracts compensates for and distracts us from the realities of continuing exploitation in the present: massive quarrying operations ripping open the land, together with a crushing plant and a continuous stream of aggregate lorries rolling through the landscape. We forget the loss of large areas of this landscape to dense and dark conifer plantations, the booming sounds of grenades and the splatter of gunfire. Through contemporary management practices we are encouraged to forget both the historic and contemporary heathland realities.

On the one hand the essentially romantic heathland that is being produced today has roots in practices of the late eighteenth and early nineteenth centuries, when landscaping mounds with picturesque stands of pine trees were created along carriage drives leading to Bicton House, which stands amid landscaped gardens on the eastern edge of the heathlands. On the other hand such a sensibility towards the landscape is combined today with a nostalgia and a feeling of melancholia for a lost landscape, of times past that were better, a sense of a lost identity in the wake of modern estrangement from the landscape and a sense of the erosive and destructive capacity of a globalized market economy. This provides moral force and value for the contemporary conservation agenda: that all this is being done for the wider good of humanity, that if we lose the heathland we are losing part of ourselves. In the words

Figure 2.12 Eighteenth-century landscaping mound, Woodbury Common, Four Firs

of Tom, the Natural England advisor, 'its about thinking about what you want your grandchildren to be able to see and experience'.

Tom's cognitive map shows seven different named administrative areas of the heathlands from north to south, marking approximate boundaries and showing roads across the heathlands from west to east and the B3081 running along the western side. Plantations, the quarry and tracks and topographical details are not shown. One prehistoric monument, Woodbury Castle, is marked, as is the golf course to the west. Within each administrative area a few characteristics and problems from a conservationist's perspective are noted, e.g. bridleway problems and bike damage. Pebble pavements and military camps are noted on Aylesbeare, as are the grenade range and model aircraft flying field on Woodbury and Colaton Raleigh Commons, and mires and nightjars on Bicton and East Budleigh Commons. The map is a succinct expression and summary of some of the key features of the heathland landscape.

A contested landscape

The heathlands are a contested landscape (Meinig 1979; Bender 1988; Bender and Weiner 1991; Tilley 2006) because there are inevitably considerable differences between particular individuals and groups with respect to the heathlands' management and use, and how and why they value them (or otherwise). Four main areas of contention are as follows:

1. Its continuous use as a military training area and its designation as an important site for nature conservation.
2. Its economic exploitation by quarrying and forestry enterprises and its preservation as a rare lowland heathland habitat.
3. Between archaeological and environmental conservation and management objectives.
4. Between the provision and promotion of public access and the conservation of the environmental and historic resource.

Beyond these main issues there are a multitude of others involving the effects, interests and values of different user groups: walkers and dog walkers, horse riders, mountain bikers, model aircraft flyers, fishing interests, and the manner in which these groups' recreational use of the heathlands relates to those who work in it and are concerned with its conservation. In the sections below we discuss issues 3 and 4 listed

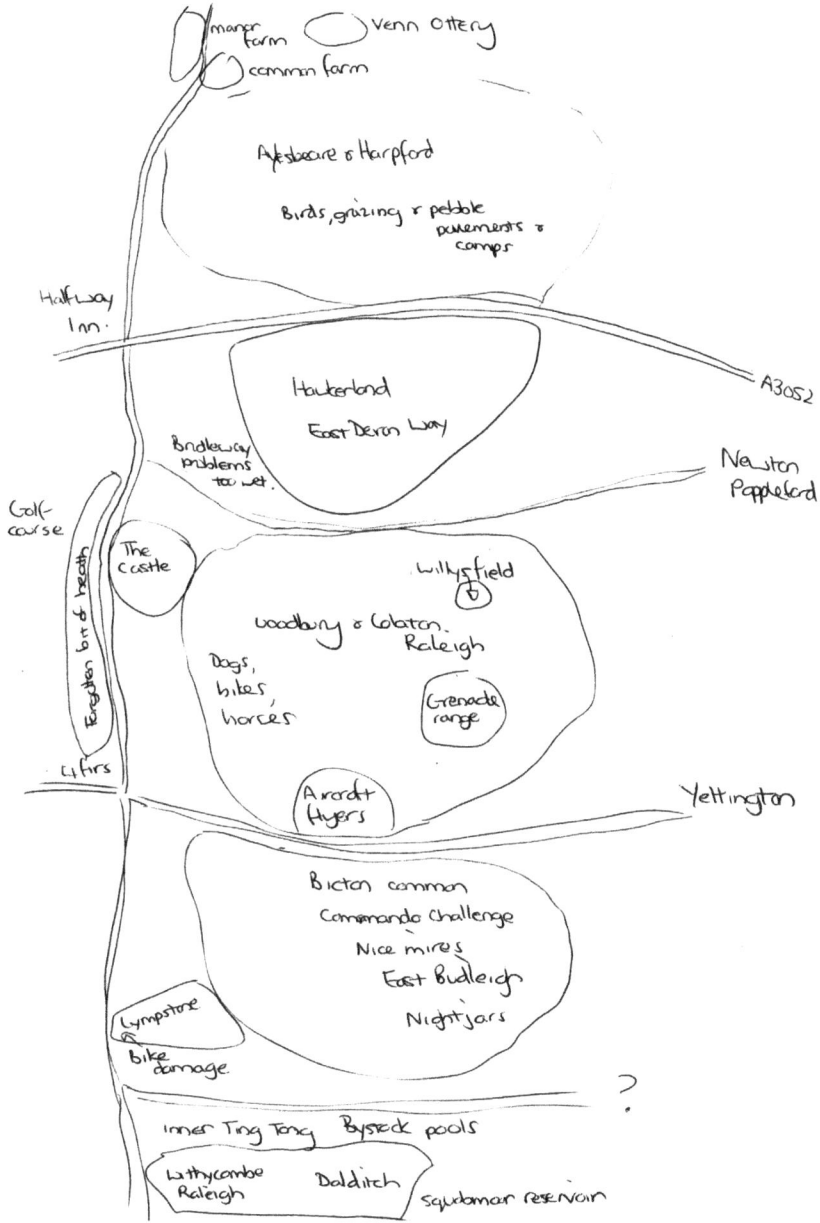

Figure 2.13 Tom's map (Natural England advisor)

above. In subsequent chapters we discuss issues 1 and 2 and the often variable and conflicting interests of recreational users.

Conflicts in conservation management

'It's English Heritage versus Natural England at the moment, you know, and we are just standing here saying, well, make your mind up, because, you know, while we are sitting here watching it, it's going as a heathland'.

(Pete Gotham)

Topsoil scraping is a new technique for heathland management which was first introduced around 1991 on the Aylesbeare and Harpford RSPB reserve. According to NE guidelines ideally 1–10% of the heathlands should have bare ground. In the RSPB reserve the amount of bare ground that has been created is probably beyond this upper limit. Scraping the soil is primarily driven by the need to achieve the target level of bare ground, something that is measurable and can be pointed to in relation to achieving the target level of what 'favourable heathland' is according to NE SSSI guidelines. The technique involves machine removal of the topsoil and accumulated nutrients in this soil. Thus a nutrient-poor habitat appropriate for heathland is maintained. It also creates temporary areas of bare ground habitat, thought appropriate for generating a mosaic of habitat types across the heathlands to encourage annual plants, lichens, invertebrates and some species of ground-nesting birds. It diversifies the age structure of the heathland plant communities across the heathlands as a whole. This scraping is supposed to remove the mat of surface vegetation while leaving the surface soil undisturbed. In practice virtually all the topsoil scraping that has taken place on the heaths has removed everything down to the bedrock, leaving extensive exposed areas of pebbles without any soil whatsoever, so it has amounted to complete turf stripping or removal.

Regeneration of the heathland habitat of heather and gorse in the dry heath areas where scraping has taken place is slow, taking ten years or longer. One of the reasons for the practice of wholesale turf stripping rather than simply removal of the topsoil humic layer is that the surface soil in most areas is very thin, often only 10 cm or less in depth; mechanical removal is very difficult, especially on sloping ground, and requires a highly skilled machine operator. The material removed has usually been dumped in long linear mounds adjacent to the scraped areas. The pebble-scraped areas and linear mounds of soil encouraging the prolific growth of the tall European gorse areas look very artificial in character and are considered unsightly by some members of the general public.

Figure 2.14 Topsoil-scraped area, Aylesbeare Common, with vegetation regrowing in basal ditch sections of circular structures

As regards the historic environment this work has the unintended consequence of destroying the archaeological resource, since flint scatters, settlement debris and shallow sub-surface features will be entirely removed. Their presence is often impossible to detect even if the area is cut or mowed or swaled prior to the scraping operations. The scraped areas include, in the RSPB reserve, unfortunately the most sensitive area of the heath from an archaeological point of view: the summit and northern slopes of Aylesbeare Common, where a series of Bronze Age cairns and pebble platforms, sites of Napoleonic date and other structures are known to exist. An archaeological survey of one topsoil-scraped area undertaken in 2008 by Tilley revealed a series of circular structures that had been almost entirely removed by topsoil scraping on a slope that in some areas had seen 50 cm or more of material removed (Figure 2.14).

What is ostensibly good for environmental conservation has proved to be very deleterious as regards the historic resource. It appears somewhat ironic that a major national conservation body (the RSPB) should have been involved in this practice. But the RSPB is still very much a single issue environmental organization, as is NE, and that is its main focus.

The linear banks and rounded mounds of scraped material, 2 m or more high and 5 m wide, create new landscape features that are inappropriate and intrusive in this open heathland landscape and detract from the landscape settings of visible archaeological monuments. Given subsequent erosion and the nature of the vegetation they are likely to be a source of considerable confusion to archaeological surveys and landscape interpretation in the future. A rolling programme of scraping, clearing new areas every year and leaving previously scraped areas to gradually regenerate, would in the long-term result in the wholesale destruction of the entire archaeological resource of the heathlands. None of the scraping operations undertaken were archaeologically monitored, nor was any information provided to the relevant local authorities concerned with the historic environment as to where and when they were to take place. This was because the heathlands are in private ownership and all but a few archaeological sites remain unscheduled, and therefore have no legal protection. A substantial change occurred, however, when the heathlands were entered into the current HLS agreement in 2006, because of a requirement that heathland management practices be sensitive to the historic resource as well as environmental conservation. Prior to this the former was hardly on the agenda. Part of the reason for this disregard for historic conservation, an attitude that still exists, is the very different conceptualization of the landscape and the value of the landscape in relation to conservation objectives held by those concerned with the heathland as a historic versus an environmental resource. Topsoil scraping obviously alters the heathland's historic character, introducing the same kind of destructive practices that occur elsewhere. There has been a failure to recognize that the entire heathland landscape, rather than tiny bits of it, needs to be protected in terms of the underlying and largely historic resource. The soil-scraping activities that have taken place to date *may* have destroyed few archaeological sites or they may not. We simply do not know anything about this, other than the fact that such areas have been destroyed for good from an archaeological point of view. The landscape has been managed for its environmental significance but with deleterious consequences for its cultural and historic significance.

As discussed above, conservation management plans currently exist and operate in terms of the natural environment and its enhancement, with substantial finances being made available. However, there is no equivalent in terms of the cultural resource, which is by contrast seriously neglected. Awareness of the problem and threat posed by topsoil scraping to the archaeology of the heathlands expanded in 2006 in the context of the new HLS agreement for its management. In July 2007 Cressida Whitton, the archaeologist for DCC responsible for the heathlands, objected to the continuance of topsoil scraping. She cited damage to previously unrecorded pebble platforms on Aylesbeare

Common, sustained during topsoil scraping operations conducted in 2006 but also recorded as long ago as 1996, soon after these scraping operations first began. Although NE has an advisor on the historic environment no internal objections or concerns about the practice appear to have been raised. A two-year moratorium on scraping was eventually agreed in July 2007, but by the spring of 2009 both NE and the RSPB wanted to resume the practice. A field meeting to discuss the issue was held on 19 March 2009 at Woodbury Castle car park.

Topsoil scraping moratorium meeting: fieldwork notes

We are meeting at Woodbury Castle car park. This is because it is the only place on the heathlands that everyone concerned with their management knows how to find easily.

Present: Tom (NE), who has convened the meeting, and another representative from NE; Toby Taylor, RSPB warden; Cressida Whitton and Bill Horner (DCC: Historic Environment Resource (HER) service); Bungy Williams (PHCT); Chris Tilley (in the role of archaeology advisor) and Kate Cameron-Daum. We stand around in a circle in the car park. Tom says the purpose is to reach some kind of consensus about future management strategies for the heathlands. We go round and introduce ourselves. 'Shall we establish first principles or go and look at specific areas?' We decide on some initial discussion and the issue of topsoil scraping is immediately raised by Cressida Whitton. Tom states that management strategy has changed so that they want to remove only a very thin humic layer in some new places on the heathlands in the future. He demonstrates the thinness involved with his thumb and forefinger. But might old areas of topsoil scraping be re-scraped in the future? The object as always is to obtain a mosaic of different vegetation types, from bare patches to newly sprouting patches to full-blown head-high gorse across the heathlands as a whole. 'What about the archaeology?' Chris Tilley says that in all probability existing scrapes have destroyed everything. There are three different issues/levels:

1. subsoil remains, e.g. pits and trenches cut into the pebbles, that might survive even in scraped areas;
2. visible structures such as pebble mounds and platforms that in all probability will be visible on the surface after close vegetation cutting (less so with swaling); and
3. flint scatters that would be expected to be found just above the pebbles, i.e. in the thin humic soil layer the managers want to remove.

A watching brief could salvage or prevent destruction of 1 and 2 but 3 is unlikely to be recognized before it is too late. It is also pointed out that the heathland managers have in the past not been topsoil scraping but conducting wholesale turf removal down to the pebble bedrock. Tom reiterates that this is now to be regarded as an abandoned practice, but they want to create bare areas for silver-studded blue butterflies (not a species specified as being important enough to have the heathlands listed as a SSSI or Special Protection Area [SPA]). Cressida Whitton and Bill Horner reiterate points about being careful, the need for a watching brief and the fact that the archaeological resource, unlike butterflies, cannot be replaced. Once it is gone, it is gone. So we agree on some principles: no fresh topsoil scraping in the old style, and archaeological watching briefs. But NE still wants to clear new areas and the main reason is now for rare butterflies.

Tom is keen that we decide something on practicalities today for areas they want to scrape or change in the new way next year. And it is Bungy who knows where these areas are. Bungy wants to take us to the south of Woodbury Castle to see some sample cleared areas and discuss more. We all get into two four-wheel-drive vehicles, and visit an area only about 300 m south of the hill fort, next to the road. Bungy says the area has been cut free of vegetation and sprayed with herbicide to kill the bracken. There are two obvious vegetational villains in this area: bracken and high gorse, where the managers want low heather. Bungy pokes at the soil with his boot: 'see how black it is?' The soil here is too deep and too rich in nutrients. Unless it gets scraped away the gorse will just grow back. Heathlands should have a thin, poor, acidic soil, and it isn't here. The problem is that this whole area is considered degraded and it is big – all the way down to Four Firs and beyond. Bungy thinks the soil is good (rather, bad) in this particular area because it has been cultivated and improved since the Iron Age. We drive back up to the castle car park and then down another track to the south-east for a much longer distance.

We come to an area unlike the previous one, where the ground has been topsoil scraped/turf stripped. The soil here is very different, red-brown rather than black in colour, and there has been very little vegetation regeneration. There is a big mound nearby composed of the removed material. Tom is pressing on whether it is OK to re-scrape such an area. Well, yes, all the archaeology has probably already been destroyed. But a watching brief might still record subsoil features such as pits.

Back into the cars and now down to the east on the south edge of Colaton Raleigh Common. Here an area next to the footpath was scraped in the 1950s and a mound of material scraped up alongside it. Bungy wants to remove this and fill a depression with the material on the eastern edge of the Commons. The reasons are to improve the habitat for butterflies, and aesthetic: opening up views to walkers, etc., rather than having the low bank there and presumably making the heathland seem unenclosed/bounded. Cressida gets out her maps to see if anything is recorded here. Just a parish boundary that may never have been marked by a bank, unlike others. She'll check but it seems OK.

Now we drive down to the Dalditch camp area in the south of the heathlands. Here there was a massive Second World War camp; now only hut foundations remain. Bungy says most of it was removed during the 1950s. We park at the car park and walk rather than drive. Bungy shows us some more places frequented by the rare silver-studded blue butterfly. Here they want to scrape. Well, there is some confusion now about the terminology – it's going to be a new kind of scraping. First the area will be cut (with an archaeological watching brief – presumably now agreed) and then the topsoil will be 'scratched' or 'agitated' with the bucket of a machine. No real damage will take place. But it is difficult to envisage how this will turn out in practice. Chris Tilley suggests that a test of this new kind of scraping would be good, to see how it actually turns out to be in practice. Cressida and Bill say they will have to check the maps to see what sites, if any, are there, and that the managers should not remove the foundations of the military buildings. During the course of the conversation Bungy keeps on saying 'she' wants this area scraped and 'she' wants that area scraped for the good of the butterflies. We wonder who 'she' is. 'She' turns out to be a local amateur butterfly enthusiast.

We cross the road to another area with military building foundations and lots of buddleia bushes. Great for butterflies, says Bungy (it is commonly known as the butterfly bush), but NE want to cut down these bushes, thus eliminating the habitat for apparently unwanted butterflies, and do scraping instead for the benefit of the silver-studded blue.

Since the meeting discussed above no new areas have been scraped in the area of the heathlands managed by the PHCT but re-scraping of previously scraped areas, now with an archaeological watching brief has taken place in the RSPB managed Aylesbeare and Harpford nature reserve, together with new areas, undertaken with a watching brief.

Topsoil scraping, a non-traditional practice of heathland management, is a fraught and emotional issue both for environmental conservationists and archaeologists and it has generated considerable uneasiness between those involved. From the point of view of the former it is considered to be an essential element in creating a desired heathland mosaic, about which the archaeologists apparently care little. For the latter it remains potentially a major and unnecessary threat to the integrity of the irreplaceable historic resource. The bare areas created and their associated mounds alter fundamental characteristics of the heathland landscape in an undesirable manner.

Swaling

Swaling or burning of the heathland vegetation is a traditional method of managing the heathlands, practised from their first creation in the Bronze Age and onwards. Gilbert White in his *Natural History of Selborne* notes that

> though (by statute 4 and 5 W. and Mary, c. 23) 'to burn on any waste between Candelmas and Midsummer, any grig, ling, heath and furze, goss or fern, is punishable with whipping and confinement in the house of correction'; yet in this forest, about March or April, according to the dryness of the season such vast heath-fires are lighted up, that they often get to a masterless head, and, catching the hedges, have sometimes communicated to the underwoods, woods and coppices, where great damage has ensued. The plea for these burnings is that, when the old coat of heath, etc., is consumed, young vegetation will sprout up, and afford much tender browse for cattle; but, where there is large old furze, the fire, following the roots, consumes the very ground; so that for hundreds of acres nothing is to be seen but smother and desolation, the whole circuit round looking like the cinders of a volcano; and the soil being quite exhausted, no traces of vegetation are to be found for years.
> (White [1788–9]; 1977: 25)

White is referring to swaling on the Hampshire heathlands. This was supposed to be restricted to the period after midsummer but often, as he notes, took place considerably earlier. This is also highly likely to have been the case on the Pebblebed heathlands in the recent historical past and earlier, the main purpose being to stimulate browse for cattle at little or no expense.

Today swaling is undertaken not to provide food for cattle but as a management technique to diversify the age structure of the heathland vegetation. It takes place in restricted areas, usually with machine mowing of the vegetation on the perimeter to act as a firebreak, with the gorse etc. in the space created being burnt off. Swaling now takes place during the autumn and winter months in order to avoid the bird-nesting season; this is a legal obligation. At this time of the year the ground is invariably saturated with water and a deep or clear burn of the vegetation is not really possible to achieve, leaving blackened gorse branches and much other surface vegetation, such as clumps of moor grass, virtually untouched:

> What you need is a double burn or a treble burn. What you need is to burn the area, get it to dry out and then reburn it, which would be virtually impossible … We're only allowed to do it September to February. And the way the laws are going the chances are we won't be able to swale in ten years' time.
>
> (Bungy Williams)

The result of contemporary swaling is very different from the kind of swaling undertaken as late as the 1930s during the summer months, which allowed George Carter, the pioneer archaeologist of the heathlands, to detect even minor surface archaeological features on Aylesbeare Common and elsewhere. White's account demonstrates the considerable dangers of uncontrolled burning but also interestingly enough shows that the effects of these fires were similar to topsoil scraping (as opposed

Figure 2.15 The effects on the heathland of a summer wildfire, 2010, Colaton Raleigh Common

to wholesale turf stripping), entirely removing the humic layer and so maintaining a poor, acidic heathland environment.

Swaling during the winter months can never produce these results:

> You never actually here get a heathland that is dry enough to burn properly whereas natural burnings would probably be in August or thereabouts when it is dry as a bone and it would take all of the heathland ... it's only the summer fires really that do an effective job of taking all the litter away as well as, you know, thereby getting rid of most of the nutrients and getting back to a proper dry heath.
> (Pete Gotham, RSPB)

However, unlike topsoil scraping and wholesale turf stripping, it is favoured by archaeologists because it aids archaeological field surveys attempting to locate field monuments (albeit far less so than traditional swaling) and still leaves sub-surface archaeological material, e.g. flint scatters, undisturbed and intact.

We can therefore surmise that turf stripping and topsoil scraping have, in part, become deemed necessary as a new management technique for producing desired areas of bare ground on the heathlands because swaling now takes place in an entirely different season from traditional practice and no longer produces the same result. This is because of the recent priority given to safeguarding the heathland habitat for nesting birds during the spring and summer months under the new SSSI and HLS priorities.

Grazing

Ever since the first formation of the heathlands around 2200BC they have been maintained primarily as a result of grazing by animals, principally cattle and sheep. These kept down encroaching scrub, through their movements creating a complex and variegated structure of plant communities of different ages and small, irregular patches of bare ground in the course of their movements. Swaling was undertaken not as a technique for conservation management but as a means of producing good grazing for animal husbandry. All the parish boundaries include areas of heathland onto which commoners with grazing rights could turn out their livestock and geese during the spring and summer months and keep them there for little or no cost. Other economic benefits to the

Figure 2.16 Woodbury Common in the 1930s. Photograph by George Carter

peasantry included furze (gorse) cutting and peat and turf cutting for fuel, and the collection of ashes to fertilize surrounding agricultural land. Old photographs of the heathland dating from the beginning of the twentieth century up until the end of the 1930s show, by comparison with today, relatively little tall gorse and heather cover over large areas. These were primarily grasslands (see Figure 2.16). Historical accounts (Vancouver 1969) indicate that the heathlands would almost certainly be described today as overgrazed and overexploited by conservationists, like other contemporary upland areas of moorland in south-west Britain, such as Bodmin Moor in Cornwall.

Today commoners and commoners' rights to grazing have been extinguished. There is only one registered commoner left, who does not exercise grazing rights. Animal grazing of the heaths and furze cutting and burning effectively stopped during the Second World War. As a result of the war there was a general agricultural abandonment of the heathlands. The landscape was rapidly transformed, with the encroachment of scrub and bracken in some areas and the development of a fairly uniform and dense cover of gorse and heather, producing the heathland landscape that we see today. After the Second World War what had once been a working part of the agricultural landscape became a landscape for leisure activities and military training, and one in which, through its designation as an SSSI and an AONB, a conservation interest arose. Management of the heaths accordingly shifted from viewing them as an economic resource to managing them for leisure and wildlife conservation. During the 1950s up until the end of the 1980s little management of the heathland vegetation took place except in the newly established RSPB Aylesbeare and Harpford nature reserve. In the absence of animal grazing other means – scrub clearance, swaling, machine cutting and scraping – had to be introduced in an attempt to prevent the area regenerating to woodland, a constant and uphill struggle.

A small herd of around thirty red Devon cattle (a traditional breed) were first introduced to limited areas of the heathland in 1990, starting in the summer months as part of the management of Aylesbeare Common. These were all wetland heath and mire areas, and the cattle were prevented from straying by temporary electric fencing with low visibility and little impact on the open character of the heathland landscape. The chief benefits are the creation of microhabitats and patches of bare ground, prevention of scrub encroachment, nutrient removal and selective control of grasses such as molinia and bracken through trampling. Such management is sustainable and can help in the conservation of traditional breeds. However it requires suitable stocking densities to be effective, along with close management of the cattle, including their removal during the winter. Studies involving the introduction of cattle to heathland elsewhere have indicated that overgrazing will be destructive of the heathland habitat and undergrazing will have little effective conservation value (Bokdam and Gleichmann 2000; Lake *et al.* 2001; Underhill-Day 2009).

Conservation grazing, heathland fencing and the consultation process for the future of the heathlands

> Grazing is a kind of key. You can maintain a heathland by cutting and burning it but it won't be in favourable condition because you don't have that essential part there which is the grazing.
>
> (Tom, NE Advisor)

Because the heathlands are common land with a right of public access under the Countryside and Rights of Way Act 2000, there is a requirement to seek consent to carry out activities that might alter their character or affect public access, such as the introduction of fencing. NE, the RSPB and the PHCT all regard the reintroduction of cattle to graze the heathlands as a whole, rather than limiting them to certain areas as is the case at present, as the key to the heathlands' successful management in the future. This would involve the erection of stock-proof fencing for their management and enclosing an open space that has never been enclosed before, thus irrevocably altering its character. One of the aspects of the heathlands that everybody, including heathland managers, likes and comments favourably on is that they remain unenclosed, thus creating a distinctive feel to this landscape.

The PHCT independently commissioned an options appraisal for the future management of the heathlands, with the report being published on the internet in 2009 (Underhill-Day 2009). One of the main purposes of the consultation report is stated to be to 'seek the views of local communities and organisations together with visitors and other stakeholders in the heaths' to seek 'a consensus on the way forward through regular communication and a shared understanding of the issues' (Underhill-Day 2009: 3). These involve the main management strategies: turf cutting, surface scraping, mowing, burning, herbicide spraying or the bruising and cutting of bracken, grazing by stock and the removal of trees and scrubs. These are all discussed at length using the technique of a SWOT analysis (strengths, weaknesses, opportunities and threats for each option), which takes up twenty pages of the report. For example, surface scraping is claimed to have six strengths, e.g. that it 'creates bare ground habitat for heathland restoration and regeneration and diversifies age structure of the heaths', and four 'opportunities', e.g. that it can be used to provide new paths and firebreaks, all regarded as positive. It also has six 'weaknesses', e.g. 'operations can be expensive', and four 'threats', e.g. that it 'can encourage visitors onto previously undisturbed heathland', regarded as negative. Each of the SWOT factors is arranged side by side in its own box. This SWOT analysis is a kind of quasi-quantitative analysis, the 'quasi' element being that none of the elements considered is quantified, weighted or ranked in importance as an 'opportunity', 'threat', etc. The analysis gives the impression of an objective and unbiased consideration yet quite clearly the factors considered could not be given equal weight in management decisions. For example one of the 'threats' regarding surface scraping is that 'machinery use can damage/remove archaeological features'. This is clearly not the same kind of 'threat' as the statement that 'machinery can cause soil compaction and erosion' or may 'encourage visitors onto previously undisturbed heath' (Underhill-Day 2009: 22).

Many of the factors in the SWOT analysis are used repetitively and occur over and over again in the various SWOT boxes devoted to the analysis of different methods. For example, the problem that the 'use of machinery uses fossil fuels' occurs six times, and the fact that machinery 'can be noisy and intrusive' occurs equally frequently. Rather banal statements such as these or e.g. 'smoke and ash can be a safety risk to traffic' or 'operations can be expensive' occur side by side and are seemingly given equal status with far more important and significant management issues such as public access, the use of the area as a military training area, grazing and topsoil scraping (see below).

The full consultation report and a summary of it were not printed and bound but made available for downloading on the Trust's website. Leaflets were handed out at various events organized by the Trust and the RSPB during Heath Week at the end of July 2009 (a week-long annual celebration of the heath, including organized events such as walks to see nightjars at dusk, pond-dipping activities for children, quarry tours, guided walks through various heathland areas led by members of the RSPB and the PHCT and others). Posters were put up in villages adjacent to the heathlands and letters sent out to local householders and organizations. These announced three open days held at three different frequently used car parks across the Commons and four drop-in days in village halls at which people could be informed about various management options and make their views known. Only one option, to do nothing in the future, was ruled out at the outset. Comments were also invited by email or to be sent by post to a member of the RSPB. It was stressed that no management decisions had been taken. The car park open days and drop-in sessions in the village halls were not all that well attended. Only 18 people visited the presentation set up in Joney's Cross car park during the entire day and only four people visited Knowle village hall. Overall the events attracted around 320 people, with the public response being described by those involved in setting up the consultation process as generally positive.

The problem with the consultation report and its presentation to the public was that it was so open ended and considered such a large number of different management techniques with their various advantages and disadvantages that it was difficult to comment at all. These included options that were never likely to be used, such as aerial herbicide spraying and sheep grazing. It was stressed that no decisions on future management had been made although the introduction of cattle grazing with permanent fencing for their management was in fact the preferred option of those conducting the consultation process. It was rather difficult to comment except in a very general way because no actual proposals and their consequences for the future of the heathland were being put forward.

The consultation framework explicitly adopted the so-called 'Common Purpose' strategy. This is a document setting out ways to agree the management of common land (Short and Hayes 2005) following acrimonious disputes about the management of heathland areas during the 1990s. These primarily involved objections to 'conservation grazing' because of fencing restricting public access and irrevocably altering the open character of heathland areas. The chief purpose of the Common Purpose strategy was an attempt to provide consultation guidelines which might defuse potential disputes from the outset by allowing full consultation with the public, transparency in the consultation process, considering different options and respecting and valuing different opinions.

Following the initial 'open-ended' consultation process the PHCT in the autumn of 2009 drew up feasibility options for the fencing of the heathlands and introducing cattle. CDE had by this time already fenced in a small area of the heathlands, Dalditch Common, which was in private ownership and not formally classified as common land, and introduced cattle. The initial plan involved fencing the perimeter area together with the construction of over thirty cattle grids on roadway access points across the heathlands. Currently most of the far edges of the heathland are already bounded in some manner, by field banks and fences. Constructing stock-proof fencing along these pre-existing boundary lines and the provision of cattle grids would therefore be relatively unproblematic as regards public access and the heathlands would retain their open character without having fencing running across them. This was the preference shared by both the Open Spaces Society and the commons managers. However this option was never presented for consultation to the general public because of road-safety objections raised by DCC Highway Authority in a series of private meetings in the spring of 2010. An alternative option considered in these private meetings was the fencing of the eastern side of the B3180 road running across the eastern side of the heathlands, thus mitigating against the potential road traffic problem caused by cattle straying onto the road, but this was considered to be too visually obtrusive for further consideration.

Following the private objections raised to the first scheme new plans were drawn up for fencing in smaller and more limited areas of the heathlands – the RSPB Aylesbeare and Harpford reserve, Hawkerland immediately to the south, and on Bicton Common – with temporary electric fenced enclosures being used elsewhere as before. This was presented to the public in a second consultation exercise that took place during the autumn of 2010, with written responses invited. These proposals, unchanged, were then presented for a final decision by the Secretary of State for the Environment in 2011: he allowed the scheme to go forward and much of the fencing has now been put up in a sympathetic manner, i.e. it is relatively low and much of it concealed from roads. The apathetic response of the general public to the final application was noteworthy. Ten formal responses were made, four were in favour, three opposed and three non-committal in character.

There are a number of issues raised by the consultation process. First, those organizing it were not independent and had a vested interest. They were all strongly in favour of the 'option' of cattle grazing with stock fencing. In the consultation process this was not stated in the first stage; this management option was simply presented as one among many others. The advantages of 'conservation grazing' were, however, consistently highlighted during the second consultation. The attendant restrictions on public access as a result of fencing and its visual effect on the open character of the heathlands were downplayed in comparison, as was the fact that some people, and many dog walkers in particular, feel uneasy in the presence of cattle however 'docile' they may be claimed to be. Some may feel disinclined to enter fenced areas with cattle, effectively preventing their access to these heathland areas.

Second, options for grazing and fencing which might have been put forward for consultation by the general public in an entirely transparent process were ruled out in advance of the second consultation, raising issues of local democracy and accountability. Boundary fencing and the provision of cattle grids work perfectly adequately on other heathland areas such as Exmoor, Dartmoor, Bodmin Moor and the New Forest, so it is curious that this option should have been ruled out by the local planning authorities. There are roads running through the other landscapes mentioned above which are equally as busy as the B3180, which runs along the western edge of the heathlands and around which the main concerns as regards traffic safety were raised. There had been a number of fatalities along this road in recent years and in general the traffic is very fast. However the provision of cattle grids and a 30 mph speed restriction and signs warning motorists of wandering animals might be argued to be a traffic calming method that would enhance rather than decrease road safety. In addition the motoring public would be made to realize, having crossed over the grids, that they were entering a very special and important area of the landscape of which many may at present be largely unaware.

Third, throughout the consultation process there appears to have been widespread public unawareness, or apathy, with relatively few people knowing about the fact that it was taking place, attending the events, or voicing concerns or opinions, raising the question of whether such a consultation exercise is effective.

Conserving the heathlands and managing people

The SWOT analysis in the Underhill-Day report on the heathlands includes not only the management of vegetation – undertaken to prevent the otherwise natural regeneration of the heathlands to woodland, and thus preserve an artificially created human landscape – but also a section devoted to the management of people and the effects of public visiting. Here four 'strengths' are listed, three 'opportunities', five 'weaknesses' and two 'threats'. Under 'strengths' it is acknowledged that 'ultimately the survival and effective management of sites such as the Pebblebed Heaths relies on public support which is strongly reinforced by the rights of access' (Underhill-Day 2009: 32). Under 'weaknesses' we are informed that 'access by the public and their dogs disturbs wildlife' while the 'threats' listed are that the public may oppose change as undesirable and delay necessary management and the 'risk that some members of the public see importance of access to land as overriding nature conservation value' (Underhill-Day 2009: 32). Under 'weaknesses' we also have the statement that 'requirements of different users may conflict'. This conflict is in fact built into the very methodology of the SWOT analysis itself without seemingly even being recognized as such. The analysis highlights but remains largely silent about the central issue of for whom and what the heathlands are actually being conserved – people or wildlife, dog walkers or Dartford warblers? – and what social and political issues are actually at stake in this management of 'unnatural nature'. At the beginning of the report the 'activities of people' are noted as being one of the main problems for managers of the heaths (Underhill-Day 2009: 4).

Curiously the SWOT analysis does not include any consideration of military training on the heathlands. There are apparently no strengths, weaknesses, opportunities or threats relating to their presence to consider, yet the 'threats' posed by dog walkers or horse riders or mountain bikers are considered serious enough to merit such an analysis. Here it can be noted that on many different occasions it is said in the report that the Dartford warbler and other rare bird species dislike human disturbance and choose to nest away from car parks and tracks, which are precisely those areas where military training regularly takes place.

Managing the heathlands from the point of view of both nature conservation and the historic environment requires managing people and their impacts. This is to a certain extent complicated by the fact that the area is common land. We will address this issue first before discussing matters of public access and impact on the heathland environment.

Originally the heathlands were common land, with commoner's rights to graze animals. Commoners' rights are normally attached to properties and are sold when a property changes hands, with the rights going to the new owner. As noted above there is only one commoner left, who has the right to graze either two horses and two cows or two horses and twelve sheep on Woodbury and Colaton Raleigh Commons, rights which are not exercised. It is something of a mystery as to what happened to the other commoners and their rights. Nobody from CDE whom we interviewed or anyone else concerned with heathland management seemed to know for certain. In other heathland areas such as the New Forest there are hundreds of commoners who vigorously maintain their grazing rights. The answer may be in the structure of land and property ownership. Much of the land surrounding the Commons is owned by CDE and managed by the estate and by tenant farmers. Many properties within easy access of the Commons that might have had rights in the past are also owned by CDE. The estate may have had an active policy of removing common rights from properties during the recent historical past. Anecdotal evidence from a few informants who said their parents or grandparents had commoners' rights suggests this may be the case, in tandem with rural depopulation, post-Second World War agricultural mechanization and intensification, changing patterns of employment and the agricultural abandonment of the heathlands, together with the failure to register any such rights under the Commons Registration Act of 1965.

In 1930 Lord Clinton signed a deed allowing the general public 'air and exercise' on the heaths. The Countryside and Rights of Way Act 2000 created a new statutory right of public access on foot across the heathland. So in principle people may walk where they like over the entire heathland. In addition the Commons Act of 2006 prohibits the construction of fences or buildings and the digging of trenches and embankments or resurfacing of the land with concrete, tarmac or other non-local material. This act effectively prevents the Royal Marines from digging new trenches as part of their training exercises (Chapter 3) and has direct implications for car park management and the erection of fencing for 'conservation grazing' as discussed above.

Dogs and dog mess

Dog walkers are by far the most numerous and most regular visitors to the heathlands, and therefore the most important user group (Chapter 6). Most come from Exmouth and other surrounding towns and villages. From a conservation perspective dogs present two major threats: they potentially disturb the wildlife, in particular ground-nesting birds, and alter the character of heathland vegetation. Dogs are supposed to be kept on a leash during the bird-nesting season from 1 March until 31 July, but hardly anybody actually does this or even seems to be aware of this. Their faeces increase unwanted nutrient levels on the heathlands, albeit in very limited areas, and furthermore are unpleasant to other users. This might appear to be a trivial issue to discuss but in fact it causes more concern and sometimes anger among conservation managers than almost any other issue:

> You know, you'll find that probably 90% of people coming round here are squeezing the shit out of their dogs … Well, of course, they only come round here because they're rich enough to be able to afford to. I mean they can walk their dog up and down their local street but there of course they have to take a plastic bag and pick it up but round here they don't do it.
>
> (Conservation Manager)

> Look at the green grass on the sides of the track. You go to every car park and there's green grass, thick green grass, on the sides of the tracks for the first 100 metres. And that is called nutrification and it is not what is required on the Commons. We want soil that has not got nutrients in it. We are proactively trying to strip nutrients off the heathland … We were quoted for ten dog bins a hundred and sixty five pounds. To clear those dog bins was £65.00 every time they were cleared. Woodbury Castle, we counted twenty-four dogs within one hour in seventeen cars. Now if those dogs all did their stuff that bin would be filled up within hours, three or four hours, so we'd be emptying that bin every three or four hours which is virtually impossible … a lot of dog people think it's their right to leave their mess everywhere. I actually had a note underneath one of those plastic bags, hanging, left on the ground and it said: 'I've done my job, you do yours'.
>
> (Bungy Williams)

> I spent half an hour on the phone with someone from NE yesterday who was giving me advice about dog pooh. It's not a strategic issue but it is an issue on the Pebblebed heaths because people take their dogs on there and they do a crap and they walk off, some people helpfully put it in plastic bags and hang it in gorse and when the team comes and strims it goes all over their hands and visors and clothes. Something has to be done or the whole place will be covered in dog pooh and dog pooh doesn't help the flora or fauna either. It changes the whole pH and fertility of the ground.
>
> (Conservation manager)

One complaint made about the public by some conservationists is that most don't give anything back. In this sense their relationship with the heathland is essentially exploitative. It is somewhere convenient and free where they can exercise their dogs, ride their bikes and their horses etc. Furthermore, they are largely unaware of the problems they cause both in terms of conservation goals and in relation to other users:

> Within the time that I've been here mountain bikes have come from nowhere, now there are mountain bikes all over the place and they cause erosion and they frighten people on horses. There is a vast increase in horse and pony riders on the common, again because its

somewhere to hack your donkey, I mean your pony, whatever it is for free, and here, because it's pebbles, well ironically you can see every time a horse puts its hoof down, it cuts the turf, or the surface, and in August especially, they loosen the surface of the tracks but because it's pebbles it rolls away. There are flints on Mutters Moor here, which interlock, pebbles roll so they create a lot of erosion … They are causing erosion; they are using it for free but they put nothing back.

(Former RSPB warden)

Beyond the issues of dogs, wildlife disturbance and erosion there are others that cannot be discussed here: problems created by littering; fly-tipping of rubbish, including materials that are dangerous and expensive to remove such as asbestos, car tyres, garden waste and abandoned vehicles; accidentally caused wild fires and deliberate arson attacks, all of which have to be dealt with by the heathland managers.

Public access and its management

All the institutions and agencies currently concerned with the heathlands and their management ultimately rely on public good will and support in one way or another. Their mantra has to be to maintain and support public access to the heathlands as an enriching experience in one form or another. On the one hand this involves providing information and facilities to increase knowledge and enjoyment of this landscape. On the other hand it requires protecting the landscape from unwanted visitor impacts. What this means, in practice, is managing what people do and where they go. While the environmental conservation objectives are reasonably well defined and 'measurable', those regarding public provision and management are less easily specified and cannot realistically be measured. The rules, principles and practices involved are to a large extent implicit rather than explicit in comparison with other conservation objectives, and there are no targets to be measured or evaluated such as e.g. decreasing or increasing visitor numbers, access to different areas of the heathlands, etc. People are a blessing insofar as public interest ultimately secures a future for the heathlands. They are also a nuisance and a problem often requiring diplomatic and surreptitious management strategies because, while many value access, few are aware of or concerned with conservation issues.

> There are a lot of tracks on the maps that are not on the ground anymore, and there are a lot of tracks that are not on the map. So it's an interesting place for map reading!
>
> (Bungy Williams)

Ordnance Survey maps of the heathlands on which visitors to the area particularly depend depict only a few registered footpaths or bridleways across them or around their perimeter edges. Some other tracks are also marked, but by no means all. There is no map available showing even the larger tracks crossing this landscape. The heathlands have in fact a complex maze of smaller and larger tracks crossing them, many of which were created by the Royal Marines and are little more than sheep tracks, difficult to find let alone follow. A common phenomenological experience is that although the heathland areas look small on the map, once one is out in it the landscape seems vast and there are relatively few distinctive orientation points. Only a very few local people who visit the heaths regularly have what might be called a practical mastery or detailed knowledge of all or most of the tracks such that they know where to go and where each track will lead. Furthermore, tracks are always changing, partly in relation to heathland management. Some grow over, others get created along the lines of freshly created mown firebreaks or in accordance with the training activities of the RM: 'the minute you open a firebreak up it becomes a walking, a different walk to somebody and within six months that firebreak is now quite a wide footpath' (Bungy Williams). To many this is a bewildering landscape in which they fear they may get lost. Consequently they tend to stick to the well-marked main tracks in the vicinity of the car parks and do not venture all that far. In our car park survey (Chapter 6) an initial idea was to provide people with copies of the OS map and get them to mark where they had gone for a walk on the map. Nobody was able to do this and show us in relation to the map where they had actually walked except in a vague manner, and so we abandoned the attempt. There is only one walking guide to the heaths, published by the PHCT. This contains five different routes across different areas of the heathland outside the RSPB reserve. There is also one mountain biking guide, produced by the Sid Valley Cycling Club, for the whole of East Devon. The latter includes two route guides across part of the

Figure 2.17 A: Tall gorse bordering footpath; B: Gorse close up

heathlands, following the major way-marked tracks. Both these guides can be purchased only at a few tourist information offices, and very few people use them or indeed know of their existence.

Although in principle people have a statutory right of access anywhere on the heathlands virtually everyone keeps to the main tracks, with the exception of the Royal Marines, simply because the dense vegetation cover of gorse, heather and bracken makes walking anywhere else very difficult and indeed arduous and uncomfortable. Areas that might encourage walking off the tracks, such as machine-cut, swaled, or topsoil-scraped areas, are deliberately obscured in the RSPB Aylesbeare and Harpford reserve by leaving a fringe of tall European gorse bordering the paths that effectively conceals these areas.

The ideal for conservation managers is that people keep to the tracks and do not wander over the rest of the heathlands, thus minimizing disturbance to wildlife and, in particular, ground-nesting birds: 'that's one of the nice things about this heath, we've got so much gorse that even bloody dogs keep off, stay on the track, you know. The gorse keeps people on the tracks and we can absorb a lot more people than on other heaths where they go off track' (Conservationist).

Another means of controlling access to the heathlands is the provision and location of car parks, since almost everyone drives rather than walks there. There are ten main car parks and some small unofficial areas where a few cars can be parked along minor roads. Four of the main car parks are located along the B3180 road following the western edge of the heathlands (Warren, Castle, Estuary View and Four Firs), a further three along the B3179 Woodbury to Yettington Road (Model Airfield, Uphams and Thorntree) and two others along a minor road to the south (Wheathill and Squabmoor). One, Joney's Cross, south of the A3052, provides access to the Aylesbeare and Harpford RSPB reserve and the Hawkerland valley area. In the recent past three other main car parking areas also existed along the B3180 but have now been closed. There is no signage along any of these roads showing where the car parks are and it is very easy to drive past them without noticing their existence. Three of the ten main car parks have elaborate, narrow, curving entrances to prevent large vehicles entering them, a response to problems posed by travellers setting up encampments in the car parks. All the tracks leading off from the car parks, and major tracks elsewhere, have barriers to prevent vehicular access to the heathlands. This is both to tackle the problem of off-road driving and fly-tipping of rubbish that has been a consistent problem. Litter picks take place every week in the main car park areas that lack litter bins, which would be very time-consuming to empty on a regular basis. People are supposed to take their litter home but many don't.

The busy B3180 and the presence of the largest and most well-known (to local people) car parks along it encourages the bulk of visitor access to the heathlands from the western edge, and it is in this western fringe that most visitors remain. Comparatively few walk right across the heathlands to the eastern side and back again or even go so far as the middle of the heathlands. On most days of the week, including Sundays, it is possible to walk in the eastern part of the heathlands and see nobody or only a very few other visitors. The same is true for the Aylesbeare and Harpford reserve:

> Certainly the Aylesbeare and Harpford reserve is quite low-key; we don't publicize it very much and that really is due to, in order to benefit the wildlife, to keep the disturbance levels down, but because we don't advertise the fact and we stay pretty low-key we have less visitor numbers [than other areas] so it's quite nice to have a refuge for wildlife within the Commons and that isn't the same in other areas.
>
> (Asst. RSPB warden)

This suits conservation managers insofar that these areas of the heathlands effectively become sanctuary areas and visitor pressure is concentrated in one area of the heathlands: the western side of Woodbury and Colaton Raleigh Commons:

When I first came here I think there were twenty-two car parks, at least twenty-two car parks, we've got it down now, and hopefully we will end up with five or six main car parks. Then you're in with a chance of being able to do something about it; about being able to control things to a certain extent.

(Former RSPB warden)

In general, publicity, and too much of it, presents a potential conservation threat for the heathland managers because it increases visitor pressure. The building of a new town at Cranbrook, a development that started in 2011 only about 8 km to the north-west of the heathlands, was mentioned on a number of occasions by conservationists as likely to cause problems in the future, particularly in relation to people trying to find somewhere nearby to walk their dogs. The heathlands are still comparatively little known and used by the public despite their close proximity to large population centres such as Exeter, Exmouth, and Sidmouth. From a conservationist's perspective this lack of knowledge is beneficial. Apart from the walking and cycling booklets mentioned above there is very little published information available about them. There is one short book published by the PHCT (Cooper 2007) available in a few local outlets, and RSPB leaflets that can be picked up at the entrance to Aylesbeare Common providing a map and brief description of the reserve. Otherwise the only information available to the public is provided on individual signboards in the main car parks. One is also provided at the northern end of Woodbury Castle, giving some information about the hill fort; another was put up by the RSPB in autumn 2011 on Aylesbeare Common, near to the Bronze Age summit barrows, but has subsequently been destroyed.

Beyond this, information about the heaths and their conservation value is provided in the annual Heath Week celebration that has been run for the last fifteen years by the PHCT, RSPB, EDCC, DCC and other local organizations, attracting around 400 people. There are also talks to local societies and at parish council meetings, and educational activities for primary school children run by the EDDC Education Ranger. The PHCT have long suggested the value of an information and interpretation centre for visitors (long planned to be sited near to Woodbury Castle but never actually realized because of lack of funding to build it), catering, bike hire facilities, etc. This would both increase public awareness and secure new income streams for heathland conservation in the manner carried out by the National Trust and other similar rural enterprises.

Conclusions

In this chapter we have discussed a series of ambiguities and tensions with regard to what a heathland landscape is supposed to be and the actual landscape that is being produced. These develop from considerable differences between an abstract conceptualization, measurement and audit management of the heathland by NE and other conservation bodies, and the heathland as produced by significant individuals and groups working on the ground. We have shown that the contemporary heathland that the public see is intimately bound up with the biographies and values of those individuals that create it rather than with abstracted notions of officialdom, although they may conform to some of their governmental strictures. Individuals can, and do, make a difference. This, we would claim, is a rather different perspective on the landscape than that espoused in some of the anthropological literature on the environment and political ecology (see Chapter 4). This often considers both the environment itself and people and their values in a highly abstracted, analytical fashion. The abstracted categories of analysis and theoretical frameworks utilized go in tandem with the reduction of individuals to bodies and 'subjectivities' considered in relation to institutions and politics (see e.g. discussions in Agrawal 2005). But our concern is with landscape rather than environment. The former is biographical, social, historical, material, specific, experiential, embodied. The latter term seems, all too often, to lend itself to abstracted, reductionist, rarefied, objectified and disembodied forms of analysis.

Another tension and ambiguity is historical: what date is the heathland as a cultural landscape supposed to represent? Is it past or present? Management strategies are constantly trying to improve the heathland and enhance it to improve biodiversity; this is very obviously a twenty-first-century agenda that occurs in relation to its transformation from a working landscape into a leisurescape. However there is still the underlying idea that the heathland should stay the same as it was in some unspecified period of the past and not change.

Contemporary wildlife conservation objectives that did not exist in the past lead to changes in techniques of management: swaling no longer becomes very effective and is replaced by topsoil scraping, which has serious consequences for the preservation of the historic resource. The introduction of cattle requires the fencing of an area that has never been enclosed, substantially altering one of the defining characteristics of the heathland: its open character. Considered in historical terms this is the single greatest transformation of this heathland landscape in its 4,000-year history.

In this chapter we have considered major issues in the management of the heathlands, conflicts between various management strategies and relations between conservation objectives and the general public in a general way. In the following chapters of Part I we consider in more detail relationships among those who work on the heath and in this landscape.

3
Bushes that move: the Royal Marines

In this chapter we discuss the relationship between the Royal Marines (hereafter RM) and the landscape. Recent discussions of the armed forces have considered their overall relationships to social and political structures, covering such issues as whether new soldiering skills are required in a 'post-modern' and globalized society in which their role has altered substantially; in a world wherein their primary objective may not simply be to defend the borders of the nation state. There has been a systematic movement towards decentralization of command, flexibility and a defence model emphasizing adaptation to a myriad of new circumstances where they must combat diffuse and unconventional forces such as al-Qaeda and the Taliban. Quality rather than quantity of forces and increasing specialization and professionalization have been key changes (e.g. Moskos *et al.* 2000; Sookermany 2011). Other studies have focused on ethical and moral issues, such as how states convince young people to go to war (Sasson-Levy 2007), issues of gender and sexuality and the construction of masculinities and their consequences: how values and norms of masculinity are structured by military training in models of military socialization, and how they in turn play their part in shaping discourses of masculinity in society as a whole (Barrett 1996; Morgan 1994; Newsinger 1997; Woodward 1998). Other studies have considered the social and psychological effects of particular military training programmes (Lande 2007; Cohen 2011; Samimian-Darash 2012) involving rites of passage and group bonding (Winslow 1999).

One of the key considerations in the anthropological literature has been the nature of the soldier's body and its social and emotional production and management. Long ago Mauss argued that bodily deportment – sitting, walking and sleeping – are all socially acquired. There is no 'natural' way in which these actions are performed (Mauss 1979 [1935]). He points out forcefully that British and French troops during the First World War were unable to use each other's spades, because these were effectively extensions of their bodies and the two groups of soldiers had acquired different skills and motor abilities.

Foucault analyses the manner in which the body is produced by the discourses and social institutions that govern it (Foucault 1977; 1978; 1980). The soldier's body becomes a surface in which power and the state become objectified, involving institutions, discourses and corporeality. Military socialization results in the internalization of norms and values that shape a new form of identity out of a civilian body, creating a different cognitive frame for engaging with and acting in the world (Frank 1991: 48–9). Frank (1991: 54–61; 69–79) links two types of bodies of the combat soldier: the disciplined body and the dominating body. The disciplined body, made predictable through training, pain and self-control, disciplines other bodies and thus becomes a dominating body. For Foucault discipline involves four main techniques: the division and arrangement and distribution of bodies, a detailed prescription of activities to be followed, a division of time into manageable segments and a network of links between bodies and their actions (Foucault 1977: 141ff.). This is of general relevance to the discussion below although it needs to be emphasized that such a perspective is highly abstract and based on 'classical' eighteenth-century French military training. It is too easily assumed that military socialization strips away individuality, producing bodies (rather than persons) all of the same sort and in the same kind of way within the contexts of different nation states. There has been virtually no study, to our knowledge, of the relationship between military training and the embodied effects of landscape in group bonding and the construction of particularized and local military identities. This is our main concern here, together with their relationship to other user groups of the heathland and environmental issues.

The RM, part of the British Royal Navy, is an elite division of the British armed services forming the marine corps and amphibious infantry of the United Kingdom. They currently have a total manpower of 7,240 personnel and a volunteer reserve force of 970. This is the largest fighting force of its type in the European Union, and the second largest in NATO. Since their creation in 1942 the Royal Marines Commandos have been active across the globe; they undertake dedicated training in the Arctic for cold-weather warfare and also elsewhere for jungle warfare. They are a highly specialized light force capable of being deployed at rapid notice. Since the Second World War all basic commando training has taken place at the Commando Training Centre (CTCRM, hereafter CTC) at Lympstone in East Devon. CTC provides training for new recruits, further specialist training in particular areas and, for officers, command training. The RM recruit training is the longest basic infantry training programme for any NATO force combat troops, taking thirty-two weeks. The East Devon Pebblebed heathlands are their principal training area in the United Kingdom. All recruits begin their training in this landscape. Throughout the training routine in the field classroom-type instruction takes place where the recruits sit on the ground in front of a trainer, followed by practical exercises in the landscape.

Recruits and the training programme

Entry to the recruit training programme follows a Potential Royal Marines Course at Lympstone lasting three days and designed to assess candidates' physical fitness and intellectual capacity for recruit training. If successful the candidate then enters the thirty-two-week training cycle. A new course of training, each time for a troop of between fifty-five and sixty recruits, starts at Lympstone about every two weeks throughout the year. Allowing for holiday periods etc. this means that the CTC trains about twenty to twenty-four troops of recruits a year, or about 1,200 to 1,400 men (there are no women). Recruits are between sixteen and thirty years old. The training involves various components, some (such as classroom training, physical fitness and weapon training) taking place primarily at the Lympstone Base and the Straight Point firing range on the coast to the west of Budleigh Salterton, others on the Pebblebed heathlands, with some more advanced training also taking place on Dartmoor, in South Wales and elsewhere.

Such is the arduousness and rigour of the training programme that there is a high attrition rate for recruits: between fifty and fifty-five per cent drop out during the thirty-two week training cycle. So less than half of those who begin their training at Lympstone will actually pass out as a RM; about 600–700 men per annum. In addition very few recruits go through the entire training regime and pass out after a continuous thirty-two-week period of training. The average time taken is about thirty-eight weeks. For some recruits it may be longer. Many have to go through part of the training cycle again, joining upcoming troops of recruits who started later in the training cycle than they did.

Each new troop of recruits is guided through the entire training period by the same group of officers, usually six in number, made up of a Captain, a Sergeant and four Corporals. They are responsible for organizing, monitoring and assessing all aspects of training. They develop an acute personal understanding of the strengths and weaknesses of the recruits in their charge. For individual practical field training exercises every recruit troop is generally divided into four sections, each with between eight and fifteen recruits, instructed by an officer.

The Woodbury Common Training Area

The RM train in all areas of the Pebblebed heathlands. Although these comprise a number of different Commons (Woodbury, Bicton, East Budleigh, etc.) the RM refer to the entire area, as do many members of the general public, as Woodbury Common. The RM training area is divided on the 1:2,500 topographic map into zones designated by capital letters (C-Z), with fire or rendezvous points (RV3, RV10, RV19, etc.) marking specific locations within the training area and the zones. These form the principal reference points for planning and organizing the training programme across the Commons by CTC. These RV locations are not known or used by the general public (except by ex-members of the RM) and represent a distinct element in a RM cartographic grid devised for their own purposes.

The RM have indelibly marked the landscape of Woodbury Common in a way that has no equivalent among any other user group. They have created their own network of tiny 'sheep tracks' across areas of the heathland where there are no other tracks.

Permanent white flag poles mark the perimeter of the grenade range danger area, with red flags raised during firing practice when access is forbidden to the general public in the vicinity. It is regularly in use on a Thursday once every two weeks. The grenade range where live firing takes place is located at the approximate centre of the main training area.

Its surrounding chain-link fence was the only area of the heathland that was permanently fenced until the recent addition of cattle fencing on Bicton Common. It also contains the only buildings on Woodbury Common apart from some upstanding remnants of the old Second World War RM Dalditch camp in the southern part of Woodbury Common. This once housed 5,000 troops before the D-Day landings in France. The booming noise made as the recruits throw the grenades at targets on the impact area can be heard right across the heathlands and beyond.

Figure 3.1 A Royal Marine sheeptrack leading up to a Bronze Age cairn used as an orientation point

Figure 3.2 The grenade range

Parts of the heathland in its vicinity, and elsewhere, are scarred with the remains of pits and trenches that have significantly altered the character of the vegetation. The endurance course, incorporating a variety of structures and natural obstacles in the south of the area, is in regular use throughout the year (see below) and forms another significant structural element of the contemporary landscape.

Except on the grenade range no live firing takes place and the RM effectively share the whole of Woodbury Common with members of the general public throughout the year. This is a very different situation from many other military training areas in the UK such as East Lulworth in Dorset, Dartmoor in Devon and Salisbury Plain in Wiltshire, from which the general public is excluded for much of the time or where the public is allowed to use designated areas or tracks only when warning flags are taken down at weekends or during holiday periods. Thus the RM, unlike many of the other UK armed services, have to undertake their training in the public spotlight.

In total the RM use Woodbury Common for 350,000 man-hours of training annually. Exercises take place during the day and at night and include map reading and orientation, signalling, camouflage and concealment, stalking, rifle training using blanks, bush survival techniques including camping and cooking, physical fitness training, ambush and combat tactics.

The RM have privileged access to the area insofar as they are the only user group who are allowed to camp out overnight (apart from fishermen around the Squabmoor reservoir) and drive lorries and other vehicles along access tracks on a regular basis. The only area that is out of bounds for military training is the Iron Age hill fort, Woodbury Castle, although even here troops may be observed passing through and picnicking on the ramparts. Today they also make very limited use of the Aylesbeare and Harpford RSPB nature reserve, in the north of the area, using it only for occasional map-reading exercises. The rest of the Commons are used intensively. Considerable planning is required in order to keep the Commons 'decongested' or to stop different troops at different stages in their training, or other groups of RM engaged in more advanced specialist training exercises, getting in each other's way.

The RM have the strong ideological support of the landowners, CDE. CDE also benefits financially from the arrangement, receiving about £15,000/annum for the lease of the land for the grenade range and another £62,000/annum for a training licence to use the rest of the area.

The RM use Woodbury Common more frequently and for far longer continuous periods than any other user group, having a twenty-four-hour presence during weekdays for much of the year. There is rarely a night when members of the RM are not using some area. Officers and recruit trainers have an intimate knowledge and familiarity with the area unrivalled by all but a few members of the general public. They regard it as 'their' landscape in both a utilitarian and emotional way. It is their taskscape, where they live and work. The heathland landscape becomes embodied in significant respects in their personal biographies and identities.

Place names and reference points

There are relatively few named places on the standard Ordnance Survey topographic maps. Some woodland areas, copses and streams lack names as do large areas of the undulating heathland topography. As a consequence the RM have their own reference points, principally the twenty rendezvous points with reference to which training is organized. For example, RV9 marks the beginning and end of the endurance course. RV14 is a Bronze Age barrow, a distinctive landmark on a high point on which a clump of pine trees grow. Other RVs mark different high points without standard map names, or car parks, junctions of tracks and roads. Gradually a recruit learns to navigate through the Commons and find out where he is in relation to these reference points. When we mentioned areas of the heathlands using names found on the ordinary topographic map, such as 'the eastern side of Colaton Raleigh Common' or 'the summit area of Aylesbeare Common', neither recruits nor trainers knew precisely which area we were referring to. Reference to an RV number elicited a very knowledgeable response by comparison.

The RM also have other kinds of names for distinctive elements of the landscape not found on standard topographic maps, e.g. Strip Wood, a long linear plantation; Split Wood, a pine plantation dissected by a road and firebreaks; Diamond Wood, a copse in the shape of a diamond. Lookout Copse is a regularly used observation post for camouflage and concealment stalking exercises. This is a square embanked tree-ring enclosure of pine trees overlooking a valley up which recruits will move. Sniper's Wood is one place where sniper training takes place and recruits camp out and so on. These place names are largely descriptive and prosaic. The Iron Age enclosure of Woodbury Castle, situated on the highest point of the Commons with its distinctive beech trees, is another major reference point regularly used in night orientation exercises. In the absence of any castle in the usual sense of this term, it is referred to as 'Castle Feature'. Through mastering these names and reference points and using them, a growing familiarity with Woodbury Common builds up, through time, on the part of the recruit, and by the end of their training none would get lost.

Bodily experience in the landscape

Apart from vehicular access and occasional helicopter drops on one specially designated zone on Woodbury Common all RM training takes place on foot. The recruit becomes familiar with the Common through sleeping out on it and developing bush survival skills and through moving across it from one place to another. The trainers and officers sleep in tents on cot beds and cook their own food. Tents are pitched adjacent to woods and copses; the recruits sleep out during the night in their sleeping bags under ponchos supported by short stakes or suspended by ropes between trees in order to keep the rain off. They cook their rations (boil-in-the-bag food) on small hexamine stoves amongst the trees. Well concealed, these are known as 'harbour' areas, from which they emerge to undertake various exercises in the training programme. Some recruits have never camped out before. Few have ever experienced wild camping. None have experienced walking in the darkness across rough terrain. Many have had little or no experience of map reading or orientation in a landscape. The initial weeks of training involve a fast learning curve relating map to terrain to body.

A crucial part of the basic training programme is to teach stealth and concealment. This involves making the body in the landscape as invisible as possible and learning techniques by which one can move from one place to another without being sensed by others. This involves not only being concealed but also not being heard by others. Controlling the body crucially means being made aware of touch and its effects. Movement is not just about controlling the body. It also involves weapons, principally the rifle. This effectively becomes an extension of the body of the recruit. The rifle needs to be kept clean and dry, ready for use at all times.

Camouflaging the body effectively means trying to make it as amorphous as possible using vegetation. Ideally it should dissolve into the landscape, become part of it. The uniform is already camouflaged in khaki, brown and green. Bracken, grass and whatever other materials are ready to hand are stuffed into the webbing covering the helmet and over the shoulders, so that they become rounded in profile, as well as into more webbing around the belt. The face is blackened with wax. In this manner the recruit will merge into the landscape and its vegetation rather than standing out and contrasting with it, figure against ground.

Recruits are taught about why things are seen in a landscape. This essentially means being aware of the significance of movement, shape, shadow and silhouette, the need to conceal light and to minimize sound. For example, recruits are trained to be aware that even such small things as the sound of a zip might be heard 200 m or more away on a still night.

Ways of moving

The 'kitten crawl' is a technique of moving when one is in close encounter with the enemy. The hands are stretched out in front of the body making sure there are no foliage or twigs that will break and make a sound. If these are present they must be cleared away. The recruit feels in front of him and sweeps away anything that is going to crack or make a noise. He then places his weight on both his elbows and raises his belly slightly above the ground so it does not drag or create friction. Using his toes he rocks forward and places his belly down again. This is an extremely slow and laborious way of moving. But done properly it is almost silent and an excellent way of moving very close to an enemy position.

The 'leopard crawl' is a standard front crawl and potentially makes a lot more noise. The weapon is cupped or cradled so as to protect it on the arms of the recruit, who moves on his elbows and knees, dragging himself forward. This is likely to make more of an audible signature but is faster. The aim is to stay low and, using cover, protect the body from being seen.

The 'baby crawl' involves moving forward using the fists and knees and keeping the weapon pointing forwards so that the enemy can be engaged at all times. Putting the fists on the ground protects the softer palms of the hands, crucial for using the rifle, from being scraped or cut. The 'monkey run' is another, quicker technique employed when there is more and higher vegetation cover; a semi-vertical form of body posture, or more of a crouch, that is rather awkward to learn. It involves standing on the back legs but moving forward in a stooped position supported by the fist of one hand, the other being used to keep the rifle pointing forwards.

The 'ghost walk' is a method of moving appropriate to jungle warfare, primarily used at night when it is difficult to see anything. The hands are used to sweep in front of the eyes to stop anything scraping the face. The foot is simultaneously used to sweep the ground and clear away anything that might break or make a sound. The recruit moves forward by placing the toes down first and then the rest of the foot. The weapon is held up ready to shoot but you look over the top of the sight in order to maintain peripheral vision. This method of holding the rifle is a lot more stressful on the body and is only possible for short periods of time. Such a walk is extremely slow and deliberate, but done properly enables silent upright movement that should not be detected. It was only when the author attempted to perform these same movements with the recruits, causing some considerable amusement and distraction for them, that their true difficulty became readily apparent. The rules of movement underlying each could only be learnt after many repeated attempts over a considerable period of time. They needed to become part of a bodily flow of movement and absorbed through the body in this manner. Thinking about what one was supposed to do and trying to move in the manner required only resulted in something clumsy and ineffectual.

Such modes of bodily movement perfectly fit Foucault's perspective in which he argues that precision and application, together with regularity, are the fundamental virtues of disciplinary time (Foucault 1977: 151): 'a sort of anatomo-chronological schema of behaviour is defined. The act is broken down into its elements; the position of the body, limbs, articulations is defined; to each movement are assigned a direction and aptitude, a duration; their order of succession is prescribed' (Foucault 1977: 152). He argues that there is a correlation between the body and the gesture. Discipline imposes the best possible relation between a gesture and the overall position of the body. The outcome of successful military training is that this becomes embodied, part of a pre-objective knowledge of the body in space-time.

The most visible form of moving across the landscape involves standing and walking forward with head held upright. The rifle is held forward with the butt in the shoulder but lowered, a position termed 'ready alert'. This is not a normal form of walking but involves moving forward while constantly swaying the body from right to left all the time. The rifle moves in tandem with the movements of the body of the recruit. This allows the recruit to constantly scan and search the terrain across which they are moving and observe any signs of the enemy either in front or to either side of him. It allows one to be observant while making a good pace. Such a mode of patrolling should be undertaken confidently. The soldier should be able to observe and to be seen by others to be observing, thus instilling confidence rather than seeming to be timid or afraid. Such 'mind games' may be of the utmost importance for success and in relation to a civilian population. Troops always move forward with one man behind the other with a distance of about twenty or thirty metres between them, never side by side. They thus significantly reduce the risk of being ambushed or decimated all at once. In these ways of moving, as the potential visibility of the recruit increases the importance of reducing sound and the significance of touch decreases.

Recruit training thus requires the development of skilled ways of moving through the landscape. It also involves learning techniques for how to cope with obstacles, whether natural such as crossing a bog or human-constructed such as ditches, barbed wire fences or walls. Battle training involves other forms of coordinated movement, such as running some distance apart down and up slopes in a zig-zag fashion so as to avoid enemy fire, with one recruit providing cover from behind for another running forward. The kind of movement that is most appropriate depends on the terrain, the vegetation cover and the objective in hand in relation to an enemy. Thus open ground can in some circumstances facilitate movement. In other situations it can be regarded as being as much an obstacle as a barbed wire fence.

Looking and seeing

The recruit thus learns to move through the landscape in a variety of different ways in relation to the perceived threat or proximity to an enemy. Whether visible on patrol or stationary, recruits are also taught to look at the landscape in a different kind of way. This might be termed a technical or functional way of looking and is about as far removed as one can get from an aesthetic, romantic gaze. Recruits learn both how to observe and how to sketch the landscape, a technique known as 'panoramic sketching' from a given point. The landscape is visually scanned and searched. This entails systematic observation from the near to the middle to the distant ground. The recruit observes that which is closest to him to the left, centre and then the right moving the head from left to right and then right to left, then left to right again and so on until he reaches the limits of his visual field on the horizon. He will try and spot anything that seems unusual or out of place, in particular the presence or potential presence of the enemy and where they might be hiding. While patrolling and moving through the landscape a similar technique will be used.

A panoramic sketch is an attempt to reproduce such a visually searched and scanned landscape from a fixed lookout point that is easily accessible to others, for example to pass from one sentry to another so that the guardsman knows where and how to look in a simple and easily portable manner that requires no specialist equipment. The recruit learns to make such a sketch by superimposing a grid over the sketch of the visual field and by using standard rules of perspective to convey the distance of an object. Features represented on the sketches include outstanding points, rivers, woods, particular trees or bushes, roads, tracks and buildings, etc. Height is represented by standard techniques of shading, and by using contour and vertical lines.

The recruit needs to be able to cope with the landscape and move through it with speed and efficiency, either by day or night. Part of this involves learning how to read and use maps, take and follow bearings. On more advanced exercises recruits will receive instructions on detailed route cards showing them where to go, from one checkpoint where trainers will be waiting for them to arrive to another. A typical route card gives the map grid reference from a starting point to a checkpoint, the magnetic bearing to be followed between the two, the distance between the two, the height as measured by contours that will be gained or lost (e.g. six contours = 60 m uphill), the number of paces that will be required to cover the distance and the expected time that it will take. A standard calculation is 4 km/hour on flat terrain. Extra minutes are added on for uphill or downhill movement. So for example moving uphill for six contours or 60 m requiring 877 paces on a bearing will mean that a recruit will be allowed an additional six minutes to move between two checkpoints 1.35 km apart. Thus bodily movement across the landscape is regulated and assessed according to a strict mathematical formula. In order to help the recruit achieve these goals a brief description of the ground will be given: 'You will set off walking up hill. You will be covering open ground. After 400 m you will see a large wood block to your right. It will remain 200 m away from you', or 'You will set off downhill. You will cross a river after 450 m, 400 m from that river you will cross a path. You will see a lone building to your front'. The production and use of these route cards show a meticulous attention to detail in the landscape on the part of the troop training team and a rationalized system of calculation with regard to the manner in which the body of the recruit should move across it from point to point. Recruits learn to count paces by using a string with small beads on it, which they can move up or down with every pace taken like a little abacus.

To a new recruit unfamiliar with the landscape, Woodbury Common appears vast and forbidding. To the officers, who know it much better, it appears small: 'When you come back on a training team you actually start to move around and think, good god, how do they shoehorn all these people into this small area' (RM Major).

The cognitive maps drawn for us by recruits and trainers were, as we expected, very different. Those drawn by recruits show only the small part of the heathland with which they were familiar during their initial training. They show part of the endurance course (see discussion below),

Figure 3.3 Recruit's map 1

the Woodbury to Yettington road and the model aircraft car park, as well as a network of tracks to the north and 'Castle Feature' (Woodbury Castle). They record wooded areas significant to the recruits because they are 'harbour' areas in which they camp out overnight.

The map drawn by the official responsible for the land management of RM training activities was very detailed indeed.

It depicts the full extent of the heathlands from Aylesbeare Common in the north to East Budleigh Common in the south. Ten RV points are shown, as are the grenade range and endurance course, with various features marked on the latter. Beyond that plantation areas are shown, the B3180 road to the west of the heathlands, the quarry, Woodbury Castle

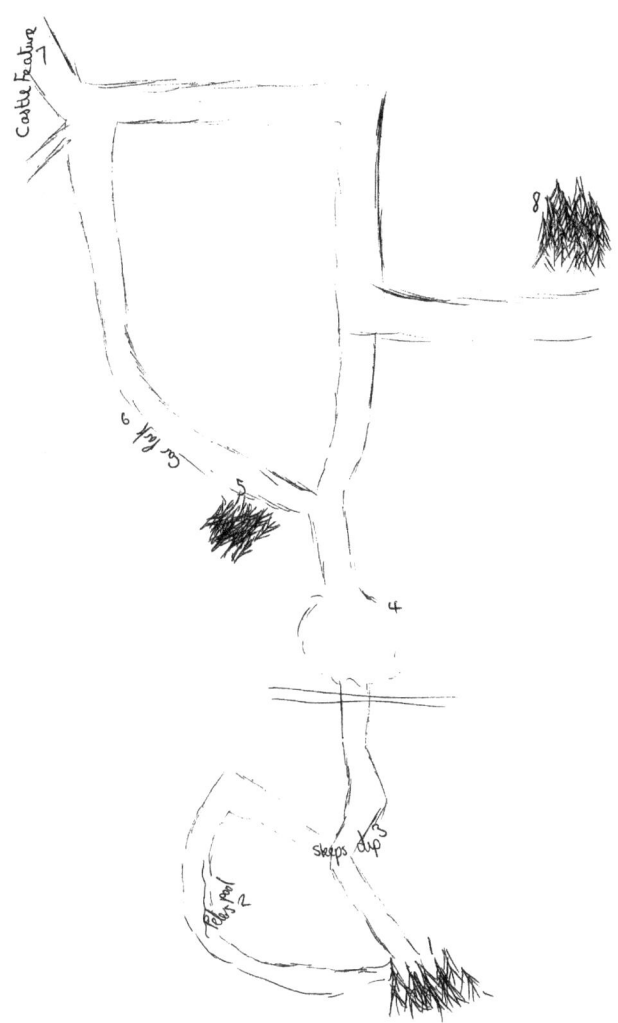

Figure 3.4 Recruit's map 2

and Four Firs, with RM names for various features given. Other maps produced by trainers emphasized features of the endurance course to the exclusion of much else (Figure 3.6), or marked it along with key features in the training routine: plantations, RV points, the grenade range, quarry and Woodbury Castle.

The knowledge of the recruit is necessarily fragmented. He is taken up to the Common for training, accessing different places for different

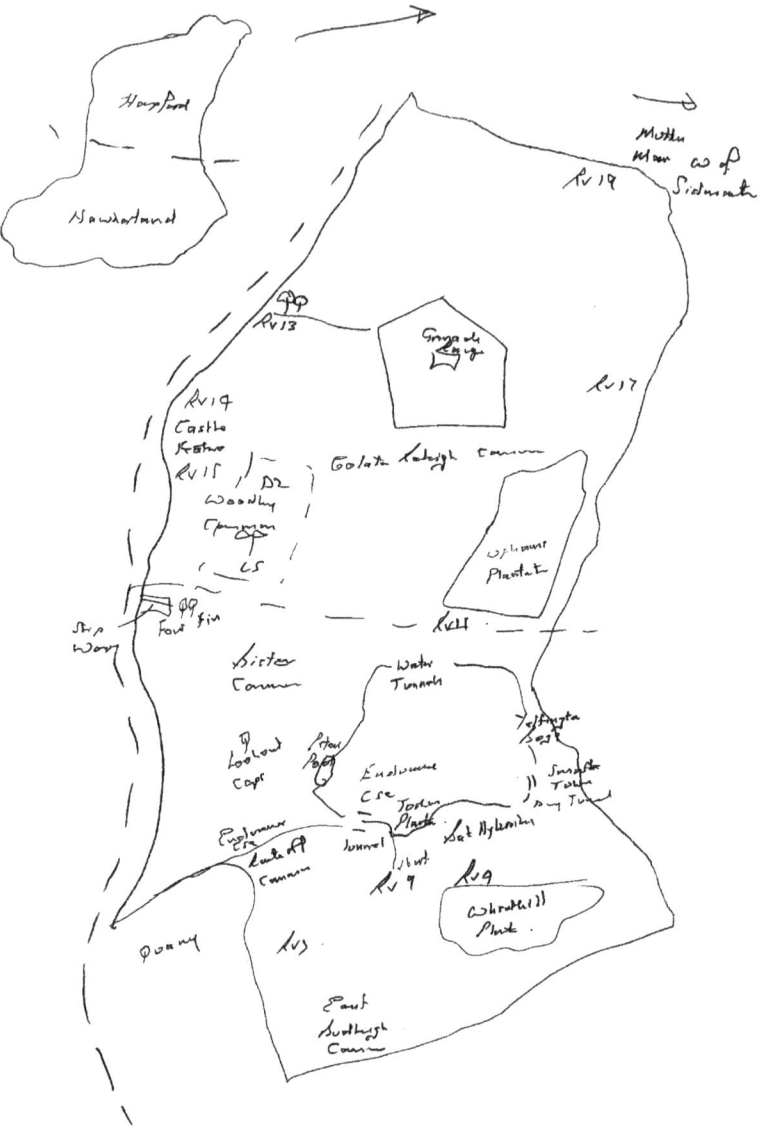

Figure 3.5 RM Lt General's map

exercises, for example to RV19 in the fourteenth week of training for a three-day exercise in which recruits are taught how to integrate a variety of skills acquired in previous field exercises. The trainers know how the different training areas fit together, the same places generally being used for the same exercises in different weeks of the training cycle over and

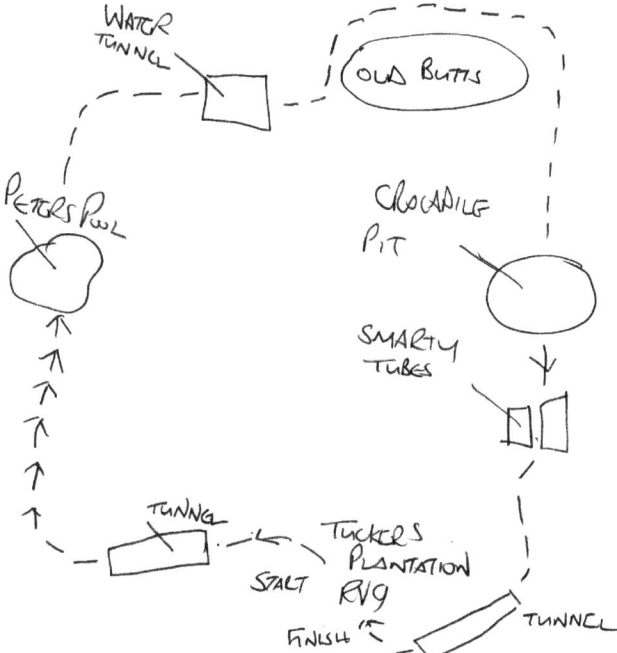

Figure 3.6 Trainer's map

over again, so for them the landscape is small. Recruits are being tested all the time and many fear the risk of failure. Heads down, performing the task at hand, they hardly have time to look or think about the landscape they are training in. They just have to move through it and endure. Trainers on the other hand can get out of their tent in the morning, have a mug of tea, and enjoy the view while the recruits are busy laying out their kit for inspection. It was the trainers rather than the recruits who thought this was sometimes a beautiful landscape to work in. For others, however, it remained simply a place of work, not somewhere where they would want to spend any of their free time.

The weapon, kit and the body

The weapon, as mentioned above, is very much an extension of the body of the recruit and an essential part of training is not only to learn how to shoot accurately at a target but to learn how to hold it, whether on patrol crossing open ground or in jungle conditions, when walking or crawling or crossing dry ground, marsh or water. It must be possible to use it at all times through being held and supported in the right way and by keeping it clean and dry. The rifle is in a sense cradled and protected from harm like a baby. An essential part of training routines involves being able to assemble and load the rifle quickly and efficiently, keep supplies of ammunition in the correct place and ready to hand and clean the rifle meticulously. The rifle has to move in a systematic way in relation to the body of the recruit, in relation to the hands that hold it, the fingers, the shoulder, the eyes and the overall posture of the body: motionless or moving, standing up or lying down, crossing water and boggy terrain or crawling through water-filled tunnels where it will be held above the head. This body–weapon relation involves a complex and ever changing articulation between the two. Movement of the body must flow into the movement of the weapon and vice versa, thus becoming one and the same. Training involves making the rifle a natural and organic part of the body that holds it. Fighting thus involves learning the motions and emotions accompanying the use of weapons. These sensorimotor practices (Warnier 2001) thus produce particular types of subjectivities and aptitudes among the recruits, allowing them to eventually pass out as a Royal Marine at the end of the training programme.

Morning kit inspection involves the recruit laying out all items in neat rows on his poncho, used as a bivouac cover at night, for inspection by the training team. This inspection covers the weapon, standard kit provided and the body of the recruit. Kit inspection covers items such as bivouac poles, flannels used for cleaning the body, water bottles, food rations and 'gash' (a RM term for rubbish) such as food wrappings from the daily ration pack. Standards of bodily hygiene are considered paramount. For example flannels used for cleaning the face or upper part of the body must be kept separated from those used for lower parts of the body. The face should be clean, without dirt or remains of camouflage materials. Visual inspection thus allows trainers to monitor the skin, the physical external state of the recruit, while examination of the gash effectively enables the trainers to assess what the recruit has drunk or eaten, to monitor what is inside his body as well. If the kit or the recruit's body is dirty or untidy, or if the manner in which they are dressed is slovenly, this may reveal much of value about their state of mind and state of preparedness for action.

The daily ration pack contains 5,000 calories, far more than the normal daily requirements for an adult male, and provides up to four meals per day. Food is considered essentially as providing fuel for the body of the recruit to be able to carry on despite the vigorous and arduous nature of much of the training. Other considerations such as taste and variety are secondary, although trainers say that taste and quality have improved over the years.

Being able to pack away all the equipment speedily and carry it in the right place, whether suspended from the belt or placed in a particular order from bottom to top in the rucksack, is another essential element in the training routine. It might take a recruit anything up to two hours to do this in initial training, a mere fifteen minutes towards the end.

Mind and body

The atmosphere in a wooded harbour area where the recruits sleep is hushed. Voices are always low; there are no lights except occasional glow lights. Use of red torches only is permitted, and no fires.

Recruits are usually so exhausted after a day of training that they go to sleep almost instantly. One of the most frequent complaints made was a lack of sleep, given that they may be woken up for night exercises or for other reasons such as sentry duty. A term used by the recruits for themselves is a 'nod'. A nod is someone liable to nod off, go to sleep, as a result of sheer physical exhaustion.

In their advertising material the slogan used for the RM is 'It's a state of mind.' This involves 'confidence, strength, independence and ability'

Figure 3.7   A harbour area

according to their website. Training transforms a civilian body into a soldier's body, and the essence of this is that it contrasts with the wrong sort of body, the kind of body that will be eliminated by the training process (Weiss 2002). All this involves a classic assertion of a mind/body dualism in which the mind is considered to be primary and separate from a machine-like body over which it will ideally exert its will. Body and mind are not considered to be necessarily in harmony or balance. Masculinity is determined primarily by a healthy body, not a healthy mind (Arkin and Dobrofsky 1978: 156). The body will inevitably give up if left to its own devices. The mind must be trained to muster up sufficient willpower to keep that body going, to counter its inevitable weakness and fallibility. Training to be a Royal Marine is thus understood by the Marines themselves not as a form of equilibrium between mind and body, but more of a battle in which the recruit is trained to counteract the 'natural weakness' of his body through mental resilience, making the body endure the rigours of training.

RM training, if successful, is thus conceived as resulting in a victory of mind over body, a classic Cartesian split. But on the other hand this may result in physical collapse or injury and the recruit dropping out:

> I know guys who have been doing speed marches who have kept on going until they have collapsed and that is the body obviously saying I haven't got any more, the mind has told me to keep going whereas if you haven't done that training your body will just step to the side and say well, no I can't do any more. It's your mind giving up before your body has given up and that's the point to push yourself a little bit more ... and it's the mind that enables you to do that not your body.
>
> (Captain and RM trainer)

A trainer is incapable of enforcing discipline from the outside, as it were. What he must try and do is to instil a regime in which the recruit learns the self-discipline necessary for survival together with differing forms of bodily deportment and movement appropriate to different situations. Above all the RM body must be capable of flexibility. Some forms of bodily movement, such as parade-ground training exercises, are mainly for the purposes of publicly visible display, moving in step, etc. Those learnt training in the landscape are of a very different character.

Being prepared for action and putting up with a lack of sleep is one element in all this. Others involve surviving physically demanding fitness regimes and induced bodily discomfort. One of the early elements introduced into the training programme is the 'wet and dry routine'. This involves wading through a bog up to the neck in water in full uniform. The recruit must then remove these clothes before going to sleep and then put them back on the following morning. All recruits regard this as trying, especially during the winter months when air and water temperatures may be below freezing. The wetness and weight of the uniform quite literally weigh down on the body of the recruit in a manner that is dispiriting. Having escaped it in order to sleep there is little that is worse than having to put it back on again in the chill light of dawn. This leads us on to consider the relationship between training, landscape and endurance.

Training, landscape and endurance

> The recruits hate this place because they've suffered up here. Give them a couple of years and they may well look back and think, well actually, compared to Afghanistan, the Commons are a lovely place. Perceptions change, don't they?
>
> (RM Lt Col)

All the RM officers interviewed considered Woodbury Common an ideal landscape for training at a basic level. Firstly the topography is varied, consisting of high points where one can look out across the landscape, ideal for developing map-reading skills. The bogs and valleys provide ideal places for concealment and stalking and add to the difficulty of negotiating one's way through the landscape and learning orientation techniques. There is a mix of vegetation from pine plantations and copses to open gorse and heather heathland. The former provide ideal harbour areas, the latter undulating rough terrain to move across by day and by night. The undulating terrain and the mosaic of different vegetation types are ideal for developing tactical skills, providing ambush and lookout points. They facilitate learning how to cross the landscape without being observed. In some areas there are useful natural orientation points such as distinctive wood blocks, prehistoric barrows capped by clumps of trees and reticulated valley systems. The landscape is criss-crossed with tracks and streams that can be integrated. In short a varied topography and vegetation is ideally suited to the basic training programme.

The nature of the bedrock – pebbles – is also beneficial insofar as it makes it exceptionally difficult to dig bunkers or hiding places, and its unstable nature adds to the difficulty of moving quietly and quickly. In the words of one recruit: 'A lot of stones on the ground. It makes you crazy running up the hills and stuff. You have to be careful not to get a broken ankle' (Recruit Jon). Recruits are taught to be careful about their foot placement, controlling the distribution of their body weights as they are going up and down hills and through the gullies on the endurance course.

In the recent past trench-digging in the pebble bedrock formed an essential element in the RM training regime. This practice was discontinued in 2002 because of objections by Natural England with regard to vegetation disturbance (see Figure 2.9). Large areas of the heathland, mainly to the south and west of the grenade range, are riddled with trenches and pits. Digging out such positions was described to us by one trainer who had himself done this as a recruit in the following way:

> I thought it was good value for recruits, good team building for four guys to dig a 4 m trench over two days. A real team builder. It means that you have to work hard. It's one of the things that toughens you up I think. It's horrendous. You can literally wear out a pick on just one trench. You had to de-turf such a large area and it is not easy stuff to de-turf and then lay your trench out and start hacking away and it would be heart-breaking to start early evening, work through the night and first thing in the morning you see what you have done, and you have only gone down a foot or two.
>
> (Sergeant, 122 Troop)

Crossing the numerous bog systems was also a nightmare for the recruits. High clumps of grass known to the RM as 'babies' heads' are interspersed across wet, boggy areas where you can sink down to your knees, making crossing them very difficult, especially at night, not only because of the wetness but also the unevenness of the terrain: 'You'll be walking and you'll stand on the side of it [a tuft of bog grass] and go over on your ankle. You know what I mean in the dark a lot of lads get broken ankles, sprained ankles' (Recruit Paul).

Large parts of Woodbury Common are covered with dense gorse that may grow up to 2 m or more in height. Gorse is an exceptionally spiky and vicious plant (see Figure 2.17). The thorns penetrate the clothing, however tough, and considerably add to the difficulties and rigours of training. All the six recruits interviewed complained bitterly about it being quite horrendous on their training exercises. The training officers, who no longer had to wade or fight their way through it, commented on its more beneficial effects in relation to toughening the body:

> I don't think I'd actually encountered gorse until I came here and obviously it's a rude awakening when you face it for the first time and you are fighting your way through it, so it is an absolutely fabulous place to conduct training because it does sort of toughen you up because it's like a constant battle moving from A to B … When you are young and naïve you just follow the bearing so if it's down deep dale and up and through thickets you just plough straight through it, so it provides a very challenging environment to conduct training and helps to bring a little bit of bite and reality into the robustness that we are trying to instil in our men.
>
> (RM Major)

'Woodbury rash' was mentioned frequently by recruits, a sometimes-chronic inflammation of the skin that might last for two weeks or more and that is caused by gorse spikes penetrating the body and causing infection. Getting rid of it might require one to lie in a bath with disinfectant. Toxins arising from infection that destroy white blood cells have been fatal in a tiny number of cases. The latest RM recruit reported to have died from this in the recent past was in 2005. Some recruits writing blogs on the RM website have been seriously worried about their health, having been infected.

Woodbury Common is thus a tough and unforgiving landscape in which to train, providing a unique combination of topographical obstacles and vegetational characteristics that make the going arduous, as does its geology. It is understandable that most recruits during training had little affection for this landscape. Sometimes it might seem like a kind of hell on earth from which one longed to escape to the comforts of the CTC Lympstone base. The officers and trainers had a rather different and broader perspective. Woodbury Common actually contributed in a beneficial sense to what the RM are supposed to be all about, a tough and resilient fighting force ready and able to cope with any landscape or conditions in which they find themselves. Other landscapes used for military training, such as Dartmoor, Sennybridge in south Wales, or Salisbury Plain were relatively easy training grounds in comparison, lacking the gorse. They were described to me as being mainly grassland areas and far less topographically varied. Their only real advantage from a military point of view was scale. They covered a far greater area, allowing

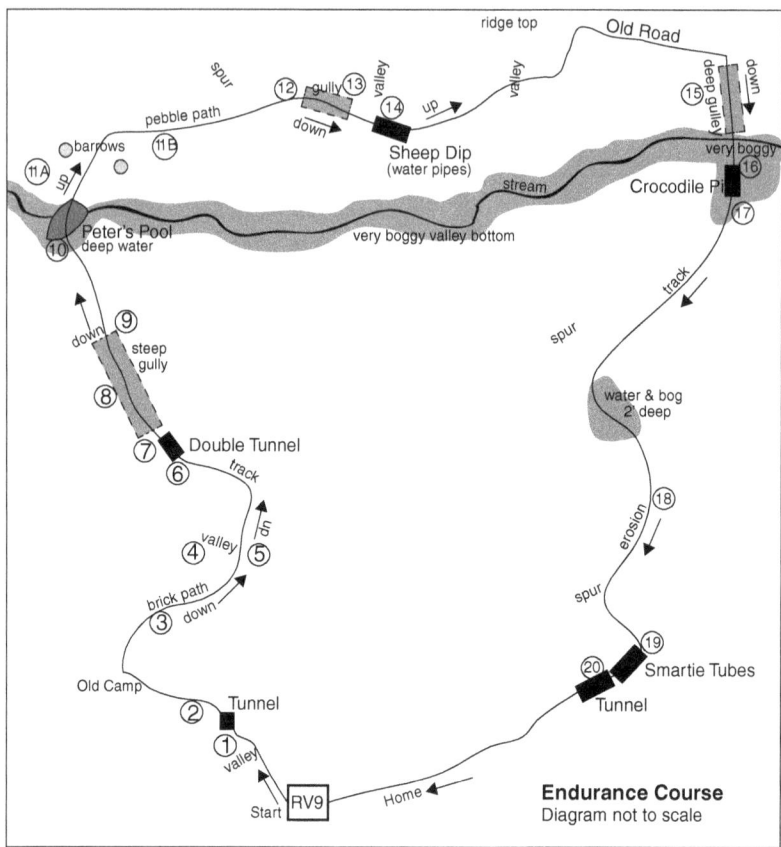

Figure 3.8 The endurance course

manoeuvres over much greater distances than was possible on Woodbury Common.

The RM were described to us by a recruit trainer as being 'substandard athletes'. They might not be able to win the marathon or excel at the high jump in the Olympics, but on passing out the level of physical fitness of a recruit would be very high indeed. To produce such bodies from what was described to me as the 'Playstation generation', not used to the outdoor life, is no mean achievement. Teaching the recruit creates superior bodily deportment. Recruits said they no longer went 'slouching around'. They felt their body was more disciplined, with a superior posture, and reckoned that you could almost certainly tell the difference between a RM in ordinary clothes and a civilian walking down the street.

The endurance course

At the end of their thirty-two-week training programme the RM recruits undergo their final tests over a continuous seven-day ritualized initiation period. These tests are the endurance course, a nine-mile speed march carrying full fighting kit; the Tarzan assault course, starting with a 'death slide' and ending with a rope climb up a 30 ft (10 m)-high vertical wall; and a thirty-mile 'yomp' across Dartmoor carrying all their own equipment. Any element that is failed must be taken again during the same week before the recruit can pass out as a Royal Marine, a fully-formed soldier.

The endurance course, the first element in this punishing schedule, is a 3.2 km obstacle course on Woodbury Common ending with a 7.2 km track and road run back to CTC at Lympstone, where the recruit must hit six out of ten shots at a 25 m target. This has to be completed in 73 minutes (71 minutes for officers). The recruits have to undertake this in full fighting kit carrying 14.5 kg of equipment, including their rifle, which has to be kept clean and dry so that they can hit the target at the end. This course was first set up in the late 1940s and the route and the types of obstacles encountered have changed relatively little since, apart from erosion creating deepening gullies and pebble exposures across bog surfaces. As a result some sections are easier now than in the past, others more difficult.

In the following section we undertake a phenomenological exploration of the course route in the landscape and its experiential bodily effects.

> It is all about that mental strength and physical robustness and that ability to endure … week 31 is when you do them. They come at the end of a 32-week training pipeline so your body has gone through that 30 weeks of debilitation, tiredness, soul-sapping way of life.
> (RM Major)

The endurance course starts at RV9 (see Figure 3.8). This is a local high point beside a minor road crossing the Common at Tuckers Plants. From here the route drops down into a wide wooded valley before leaving it, gradually ascending on a narrow pebbly track through dense gorse bushes along the flank of a spur. At the top of this path the first 'dry' tunnel, is encountered, 24 m long. This requires the recruit to bend low before entering it. The tunnel lined by corrugated iron sheeting with wooden supports reduces significantly in height, bends to the left, requiring crawling through on the knees in the darkness over the pebbles. It then curves to the left again at the far exit. The bends not only have the effect of making it dark but also of giving a distorted sense of its length.

Figure 3.9A Exit of dry tunnel

Figure 3.9B Gully

On emerging the recruit runs up a gentle slope, now along a well-defined pathway which is distinctly less pebbly and more sandy and gritty, going through remains of part of the old Second World War Dalditch camp. The recruit now arrives at a broad open space and takes a sharp right turn, now running over a coarse path surfaced with bricks from the old camp that is relatively easy to run on for about 100 m. The path drops down into a shallow, broad valley, passing the eastern end of a Second World War rifle butt and then bending to the left before reaching a valley bottom. It then rises up a short, steep slope covered with pebbles. At the top the recruit turns sharp left and proceeds for 100 m before veering right to a double tunnel obstacle built above ground, with a wooden façade, corrugated iron sides and an earth and pebble covering. These tunnels, bending to the right, again through which one must crawl, are frequently wet inside during the winter months. They have a rough pebbly surface and are 20 m in length and 0.8 m high.

The exit is on a high point by Sniper's Copse with a view down into a deep, broad valley. After 15 m the course drops very steeply down into this valley through a very deep and narrow kinked gully, effectively a roofless tunnel. This has jagged, crumbling, red pebbly sides and an unstable, rolling pebbly base. Trying to run down this is like running on ball bearings. The gully appears to be natural but in fact is the result

Figure 3.9C   Peter's pool

of human erosion, marking the passage of so many generations of RM troops, deepening every year.

There is nothing like it anywhere else on Woodbury Common except along two further sections of the endurance course. There is a fan of loose pebbles at the bottom of the gully where one reaches the edge of a broad, flat and very boggy valley bottom through which the Budleigh Brook flows east.

> There is a bog down there and Peter's Pool. You'll go up to your chest. In this weather [-2 degrees Celsius, with a strong wind blowing making it feel like –8 C] it's not going to be very nice!
>
> (Recruit Daniel)

> I went through training '79 … It was just, to me, it was just a horrible area to be dragged through, up to your neck in water, you know.
>
> (Supervisor, Straight Point Firing Range)

The course crosses the valley through a deep pool, known to the RM as Peter's Pool.

This is where the wet and dry routines take place early on in the training programme. Taller recruits are immersed up to their necks. Shorter recruits effectively have to bounce across the bottom to keep their heads above water. One arm and hand, holding the rifle, is held above the head in order to keep the rifle dry, the other is used to grasp a rope suspended across the pool to aid the recruits in pulling themselves through to the other end. The pool is almost 50 m in length, irregular in shape and up to 10 m wide, filled by rivulets of water that flow into it from the bog, predominantly from the west and north. The long axis of the pool is across the valley bottom from south to north with a straggling line of birch trees on the eastern side. This pool has been artificially modified and deepened over the years. In 2010 gorse bales were flown here and dumped by helicopter in an exercise by the RM to prevent massive erosion of the boggy ground on its eastern side, where it is possible to sink swiftly down to the knees and beyond. All RM know and remember this place and there are a number of different dits (stories) about how it acquired its name (e.g. after a frothy-mouthed ranting sergeant, or a dead recruit), but nobody knows for sure.

The course now winds and ascends through gorse. The path is only about 50 cm wide, worn down through peat sediments to a pebble base, along which a rivulet flows. After about 50 m the path emerges from the gorse onto a dry, broad shoulder of land forming the north side of the

Figure 3.9D Pebble path

Figure 3.9E The sheep dip

valley and passes between two small prehistoric barrows to the left and the right.

The recruit, now wearing completely sodden clothing that clings to, chills, and further weighs down his body, ascends a steep slope on a path now up to 2 m wide and covered with loose pebbles.

At the top of the slope the path bends to the right (east) and then descends steeply through another narrow gully that cuts through the steep edge of the valley and is up to 3 m deep. This has a loose, unstable pebble floor and crumbling, jagged pebble sides in a red sandy matrix that can easily rasp the hands should one slip. Again there is a large fan of eroded pebbles spilling out into the valley floor at the bottom. The path, eroded through the peat, then crosses a side valley to the left through boggy ground in which a steam flows. The recruit now confronts the next obstacle, the water tunnels, called by some recruits the 'sheep dip'.

This consists of three irregular water tanks constructed from concrete blocks. The first chamber collects the water from the stream running through the side valley, and this then fills the second and third chambers. These are connected by two water-filled concrete pipes 70 cm wide and 2 m long, through which, fully submerged, the recruits have to pass. This requires the teamwork of three recruits. One jumps into the first water tank, placing his rifle on top of the water pipes. Another jumps in behind him and a third into the last tank. The first recruit takes a deep breath and enters the water pipe. He is pushed and pulled through by the two other recruits in front of and behind him, who then take their turns.

Emerging from the pipes the recruits retrieve their rifles and run up a loose, pebbly slope to the north-east, ascending over the face of an old brick-built Second World War rifle butt. The course then drops down into a shallow, dry valley before rising again up a very pebbly slope. The path then proceeds east along the top of a spur running parallel to the valley to the south for about 20 m to a pine clump. Here it bends sharply to the right down to the valley bottom. The course cuts through the slope, through an extremely constricted pebble-filled gully in places only 30 cm wide at the base, scarcely allowing one to put one foot in front of the other. It plunges down through thick, gorse-covered slopes, meandering through the pebble deposits on either side up to 3 m in height. The base of this gully is filled with large eroded pebbles, some the size of boulders. It has a spiky gorse roof. The recruit emerges from the gully onto a kind of pebble causeway spilling out onto the valley floor and then crosses the very boggy valley bottom, eroded down to its pebble base. At the bottom

Figure 3.9F Crocodile pit

the recruit passes over a fast-flowing stream about 2 m wide and then enters, on the other side, the next obstacle, the 'crocodile pit'.

This is a 32 m-long water-filled and extremely muddy trench branching off laterally from the stream that fills it, crossing the bog of the valley side to the south: 'It's really, really muddy, well thick mud, which adds more weight to you' (recruit Paul). The main difficulty here is not the water but the thick, red, cloying mud sticking to and weighing down the boots and the lower legs. The RM has protected this area, like Peter's Pool, from erosion by the dumping of gorse bales next to the stream. Emerging from this the recruit clambers up a wooden ramp, entering a woodland area.

The path now winds up a slope through the woodland, along a broad dirt track that is easy to run along. It then diverges to the left. The recruit must now follow the course of a muddy, water-filled depression about 0.5 m deep, skirting another boggy area (trainers, as elsewhere, walk along relatively dry ground beside it). Beyond this water-filled feature (about 30 m long) the path continues along the margins of the bog but is eroded through the dark, peaty soil to the pebble base, meandering through clumps of moor grass. After about 200 m the course bears left and rises up and out of the valley over a small, low spur.

The recruit now arrives at the 'Smartie Tubes', named after the popular sweets packaged in long, thin cardboard tubes. These are two parallel, dark, partially water-filled pipes, dirty and foul smelling, 70 cm in diameter and 20 m long. It is easy to get stuck in them, particularly with kit slung around the belt. The recruit, carefully keeping his weapon away from the water, must drag himself through these on the belly, grunting as he moves; a lot of the recruits will be vomiting.

> Totally pitch black … You've got to work your way through them all the way through to the end and if you're claustrophobic … and it's all a race … all the water's splashing around you, you've got to keep your head high, so as to not take in the water … and you've just got to get through there as quick as you can; all you are thinking about is time.
>
> (Recruit Paul)

Immediately after exiting they stand up, only to have to immediately stoop down and enter a single rectangular dry tunnel with corrugated iron sides, whose entrance is 70 cm wide and 1 m high with a wooden façade. This tunnel, 20 m long, is bent and dark. It lowers from the entrance, swinging around to the left at both ends. On exiting, the final stretch of the course continues through pine trees, rising up a slope to RV9.

The recruit then picks up his cap comforter, the hat worn through training, from where he has left it on the ground, runs back to CTC at Lympstone, mercifully largely down-slope from RV9, and, in a state of exhaustion, fires at the target and hopefully hits it six times – all within the 73 allotted minutes.

> At the end of it when you are knackered and your knees are killing you and you have ripped your hands apart you hope you can actually fire your weapon and hit the target.
>
> (RM Lt Col)

The recruit is hardly aware of the landscape through which he periodically runs, crawls, wades or completely submerges himself in water. He moves through rather than sees it. Constantly aware of the clock ticking he has no time for any aesthetic consideration of the landscape, history, archaeology or wildlife. His eyes must be down most of the time because of the unstable pebble surfaces and boggy ground which he has to cross. When he looks ahead is when he is crawling through the dark tunnels, desperate to get through them somehow and emerge into the light at the end. He is both water-sodden and covered with mud; his clothes and kit, swung around the belt, weigh increasingly heavily and drag his body down. He must at all times take care of his weapon, protect it from harm, while being unable to protect his own body from physical assault. The recruit must be totally absorbed in the task and his body in motion suffers a kind of sensory amnesia: 'Personally you just get there, you get broken' (Recruit James).

The whole point of this training is to try and remain unaffected by the different sensory experiences to which he subjects himself and his body, to any thoughts outside the task. The endurance course thus encourages and instils a single-mindedness oblivious to either the landscape through which he moves or specific sensory experiences encountered along the way: constriction, darkness, unstable and slippery surfaces, ground that moves below you and sucks you down, wetness, cold, pain and exhaustion. The body must become a kind of machine in which ordinary sensory experience becomes dehumanized, and the experience of landscape becomes disembodied. This fits perfectly, in an extreme form, Leder's (1980) general argument that the body automatically 'fades' from conscious experience when people are engaged in purposeful activity. It is both present and absent, but in this case pain constantly reminds the recruit of its existence. Another way to understand this is in terms of Csikszentmihaly's (1990) concept of 'flow'. This is a state of consciousness in which a person is completely absorbed in an activity so that they forget themselves within it. Action and awareness become merged, distractions are excluded, self-consciousness disappears, sense of time becomes distorted and the activity becomes an end in itself with a merging of awareness into it rather than the surroundings.

The endurance course becomes, for the recruit, a kind of liminal, betwixt-and-between rite of passage in two senses. It is highly significant that all of the RM we asked to draw cognitive maps for us of the heathland depict the endurance course in varying degrees of detail. In particular it is the worst bits of the course – Peter's Pool, the 'sheep dip' or water tunnels and the crocodile pit – that stick in the mind and that are shown (see Figures 3.3–3.6).

The endurance course is a test, which the recruit must undertake and succeed at in order to pass out at the end as a RM. In order to do this, he must anaesthetize his body, try as much as possible to disconnect it from the landscape through which he moves and from the sensory assaults on his body; the less he feels or experiences anything the better. This is in direct contrast to other aspects of RM training which all emphasize – for example, in camouflage and concealment and stalking tactics – sensory awareness of an involvement in the landscape. A well-concealed recruit must instead attempt to become part of it, immerse himself in its particularity.

Undertaking the task involves aspects of cooperation and teamwork and competition between recruits, like most other aspects of RM training. Recruits need to help each other out, encourage each other to go on, through obstacles such as the water pipes. Each recruit also wants to excel personally, to do the best he can, not to fail the test because of others.

Conservation issues

> I've been to my first East Devon Pebblebeds Conservation committee meeting and I was like the Emperor in new clothes, the little boy in the back, and I said 'who decides what is natural then?' And you could see the look on their faces with NE looking slightly askance at me … All I'd say is that our impacts on the ecology is minimized, responsible, and on occasions quite positive in terms of the environment.
>
> (RM Lt Col)

On some days on Woodbury Common there is a rather different kind of dawn chorus than one might expect in a wildlife conservation area: rapid and repeated gunfire that may last for a considerable period of time, followed by an eerie silence. This may similarly occur at dusk as these are the favoured times for RM tactical battlefield exercises, attacking an enemy position at a time when it will be least expected. Sometimes what you thought were bushes start to move. Unexpectedly gunfire and smoke can break out from a copse, troops emerge and start to run across the landscape at any time during the day. In the dead of night suddenly the entire landscape will be brilliantly illuminated by high-flying flares descending gradually to the ground on their small parachutes while shadowy figures can be observed crossing the landscape. These are some of the day-to-day realities of military training visible to the general public.

All RM recruits are informed that Woodbury Common is an important landscape from the point of view of nature conservation and that they should respect it and not leave litter lying around. They usually receive a lecture about this at CTC at an early stage in their training cycle by the Commons Warden, Bungy Williams, (himself an ex-Royal Marine, who regularly uses the RV terminology to refer to specific places in the landscape), and the official who manages the land area for Ministry of Defence Estates. In general the RM are extremely sensitive to public relations issues and do their utmost to maintain a good image of themselves and the manner in which they use Woodbury Common for training. They are well aware that as an organized and highly visible presence they constitute an easy and ready target for potential criticism compared with other user groups on the Commons. Their efforts are largely successful in that very few members of the general public are openly hostile to their presence; the vast majority readily accept that 'they need somewhere to train' and are generally supportive. Only a few individuals we talked to were critical and wished they were gone. Apart from their general presence these criticisms related to noise frightening dogs and litter being left around. The RM are absolutely scrupulous about removing litter and, apart from the absence of ground vegetation in woodland areas where they camp out, they usually leave no trace behind. They even rotate their harbour areas in woodland on a regular basis in order to minimize their effect on the environmentally sustainable forestry policy of CDE.

Most litter is left behind accidentally, for example a recruit packing up quickly who leaves a box of bullet cartridges behind. There is also litter that cannot be removed because it is concealed in the dense heathland vegetation: remains of flares or smoke cartridges, or spent bullet casings. This material, concealed and invisible as it generally is, is to be found everywhere across Woodbury Common. The RM perform litter picks across the Commons on a regular basis and the majority of the material they find has been dropped by members of the general public rather than themselves: 'when we did the rubbish sweep just before Easter [2010, using 100 recruits] there were sixty-eight bin liners with trash; all but two had civilian rubbish in' (Lt Col).

Heath wildfires have been caused by the RM in the past and still occur today. As a consequence there is now a pyrotechnic ban during the summer months when the ground is dry. In general the RM feel responsible for the area and can get upset when they see members of the general public trashing it, lighting fires and dropping litter. Since they are often crawling across the ground a particularly unpleasant problem for them is dog mess left on the ground in the vicinity of car park areas by the general public:

> I mean I've got no problem with the locals using it and that, obviously it's for everybody but, you know, a lot of people do abuse it. Use if for somewhere to take their dog, you know, for a walk and

do their business and they just don't clear it up after themselves, they're a big hazard I think.

(George, Trainer)

They [the public] are dead quick to pick up brass and pyrotechnics and complain to us but they are not so keen on picking up the dog shit. It's more dangerous all the diseases that are in the dog mess.

(Trainer and Sergeant)

The almost continuous presence of the RM on Woodbury Common has other potential benefits in that it deters fly tipping (although this still occasionally takes place) and off-road driving by four-wheel-drive vehicles and BMX motorbikes. Similarly deliberate arson attacks can be spotted or prevented and wildfires reported and more speedily put out though cooperation between the RM, the fire brigade and CDE. The RM are keen to emphasize these benefits and other environmentally friendly factors relating to their use of Woodbury Common, contrasting with the manner in which the public more regularly abuse heathland areas in other parts of the UK that lack a military presence. Some members of the general public, particularly women, say they feel safe walking on the Commons because of the RM presence. Since it is only a few kilometres distant from the CTC base at Lympstone its use keeps the RM's carbon footprint relatively low. If they were not able to use the heathlands for training they would have to go much farther afield to Dartmoor or other military training areas on a much more frequent basis. Beyond this the presence of the CTC base at Lympstone greatly benefits the local economy.

On the other hand the RM's effects on wildlife in general and endangered species in particular is unknown. There can be little doubt that their presence during the day and the night, the movements of recruits across the landscape, the noise and disturbance caused by firefights and mock battles, grenades, flares, smoke canisters and so on must be detrimental to a greater or lesser extent. This is so especially in relation to the two rare and endangered species of nesting birds likely to be most disturbed, the nightjar and the Dartford warbler, the presence of which has led to the SSSI and SPA designations. Members of the general public – walkers, horse riders, cyclists, etc. – seldom cross the Common during the night. They invariably keep to the major tracks and paths. The RM, by contrast, typically utilize other areas off the tracks. Indeed, for tactical reasons, the RM generally avoid tracks, given that in a combat situation it is precisely such routes that the enemy might expect you to move along, places most vulnerable to attack. Recruits following bearings from one checkpoint to another try as far as possible to follow a straight line and carry on regardless of whatever lies in front of them. During their training exercises what is uppermost in the mind of the recruit is succeeding at the task in hand. They are not thinking about endangered bird species or conservation at all. They necessarily concentrate on trying, for example, to follow the bearing given to them and arriving at the checkpoint in good time, or stalking from A to B without being observed. When explicitly asked about the wildlife a typical response is that they 'don't really pay attention to it; it's the job at hand for us … you don't really have time to look at what's going on around you'. Half the recruits interviewed at the Straight Point firing range and those training on Woodbury Common said they knew nothing about wildlife or endangered species on Woodbury Common.

It could be argued that disturbance caused by the general public is more or less limited to areas around car parks, major tracks and paths, while the RM potentially disturb the wildlife everywhere else. In this respect it is interesting to note the very limited use by the RM of the Aylesbeare and Harpford Common RSPB nature reserve for training. One can only surmise that this may be driven by RSPB and NE perceptions of its potentially deleterious effects. It should be noted however that official members of these and other conservation bodies are extremely reticent about making any critical comments with regard to the RM. Some, indeed, prefer to play up the positive deterrent benefits of their presence, as discussed above, against the deleterious effects of the activities of some members of the general public.

Some volunteer environmentalists have a far more critical perspective. Some speak of the Marines as being part of the 'community', part of the landscape, but two environmentalists who are not connected with the RSPB made clear their not-so-positive feelings regarding the military use of the landscape:

> You'd just be wandering along, thinking your thoughts, looking at what's going on. Suddenly a grenade would go off and, oh, I really resented the whole militaristic idea. It was entering my head, you know, a grenade would go off and I would enter into all sorts of reveries about that and things, and I thought that was a very interesting thing, you know, something happens and then your whole thoughts are skewed in a different way. You arrive there, in a natural sort of place, in which birds are singing and it's all very peaceful and suddenly you get a military intrusion, and it does affect you, you know.

> 'What's happening there?' and the whole sort of militaristic comes into you. I used to resent that actually, at times, you know, when you're up there for a bit of peace and quiet and suddenly a crump from the grenade range would go up and it would intrude into your peace and quiet if you're up there for contemplative walking.
> (First environmentalist)

This particular environmentalist views the heathland as a natural place and so does the other interviewee:

> Yes, I find it a natural place except for, obviously, small pockets where you've got the quarry, and where the military charge around as well; I can never understand that but there you are ... What I object to is you have the Military of Defence, defence against what? Why not Ministry of Interfering in Foreign Countries?
> (Second environmentalist)

He acknowledges that a military presence in many areas in the UK has stopped the spread of suburbia and that the wildlife is often 'absolutely superb'. His objection in this instance appears then to be more political than concerned with potential damage or noise. As has been said, many others appreciate the presence of the Marines. One reason given is that their manoeuvres help keep the heathland's vegetation down, which aids in the construction of an appropriate habitat for certain wildlife:

> It's a bit of disturbance but essentially it's just human footfall and they're keeping the paths open, they're scuffing up bits of bare earth. So, you know, as long as it's not in the same area every single day of the year, come rain or shine, it's going to have a small impact.
> (Commons warden)

Like other users of the heathlands, the majority of the environmentalists feel the Marine presence offers the heathland protection: 'The Royal Marines go up there and they're the nearest thing to a police force; they're a presence, an authority, with much more presence than the wardens could give and they don't really do any damage' (Environmental volunteer).

It might be the case that the RM has little or only a very limited impact on endangered species. In this respect it can be noted that these species are still present despite seventy years of continuous training by the RM Commandos and a long military use of the Commons going back to the Napoleonic wars and earlier. In one of the most intensively used training areas, East Budleigh Common, nightjars are in fact more frequent than in other areas. What is entirely unknown is whether if the RM ceased to use the area bird populations would significantly increase, or remain the same, or perhaps even decline as a result of an inevitable increase in the deleterious effects of irresponsible members of the general public. The historical presence of these endangered species is also unknown: were they present when the heathlands were intensively utilized for cattle and sheep grazing, furze or gorse cutting and peat extraction up until the 1940s? Or have they colonized the area in tandem with the RM and a drastic decline in economic exploitation which has allowed the gorse and heather to grow largely unhindered since the Second World War?

Landscape, embodiment and memory

> When we leave here we all have memories so wherever we go we remember different parts of Woodbury Common and that's a good thing. We're all RM together and we remember Woodbury Common and when we leave the Core and we go out into the wider community we still remember Woodbury Common and the fun we had and the challenging sort of environment we soldiered in, so we are always spinning dits [stories] about 'when I did training I remember when we did this exercise' and Castle Feature, so people will talk about them in the Marine community. So wherever you are, if you are living in Australia or Spain, you will talk about Woodbury Common because it's a common thread to everyone since World War II.
>
> (RM Major)

Some officers and trainers said they tended to block out the bad memories of this landscape and only tended to remember the good times or amusing incidents:

> None of it was really enjoyable because when you were up there you were under immense pressure; you were being tested and you knew that if you didn't pass what you were doing at the time, you, you know it would affect you and you'd get, you know, removed from the training and put back.
>
> (George, Trainer)

Whether enjoyable or not, the entire experience of training creates strong affective bonds between troops of recruits. They have common experiences and common memories of the landscape of Woodbury Common. This is what makes them distinctive from other parts of the British armed forces and gives them a unique identity. Woodward (1998), in a discussion of the UK military as a whole, asserts a generalized 'rural' identity for military people, as manifested in the manner in which they present themselves. This effectively ignores the particularities and effects of the different landscapes in which training takes place. Formative experiences are always particularly strong and for most recruits this is the first time they have lived away from home and encountered a landscape in which they must learn to live and endure. Weather may contribute significantly to the kinds of memories the RM have of the landscape and the stories they tell. While the general public avoid the Common during bad weather the RM do not have this option. A recruit who begins training in the autumn will have a significantly different experience than one who begins in the spring. The latter will do much of their training during the summer months in which physical conditions are much more benign on exercises such as the wet and dry routine and while navigating across the landscape. In winter Woodbury Common, even though it is near to urban centres, can feel incredibly bleak, inhospitable and remote.

Since all RM Commandos have trained on Woodbury Common, carrying out similar kinds of exercises in different parts of the heathlands, the experience of this landscape creates a wider common bond between all members of the RM, whether or not they have trained together in the same recruit troop. It forms a fundamental element in their personal biographies. Thus this particular landscape is key to social bonding and their identities. It is an active force that maintains their difference from other UK fighting forces, and far more important than their uniforms and green berets in establishing whom they are.

RM can thus share 'dits' about their experiences in their landscape wherever they might be on active service: Afghanistan, Iraq, the Arctic or Sierra Leone. Memories of landscape get passed down from generation to generation in the form of these dits. These shared memories are particularly powerful amongst those who have trained together in the same recruit troop: 'We talk about Woodbury Common, Woodbury rash, crawling through the gorse, picking bits of gorse from your legs a few weeks later that are still emerging, you talk about that, Peter's Pool and the endurance course' (Captain and Trainer). They can all refer to the same named places, the grenade range and features of the endurance course.

These form the common memory of the landscape on the part of officers, shared reference points in their understanding of their taskscapes.

The RM not only share these place names but also those of named individual exercises taking place in different parts of the landscape, e.g. exercise Marshall Star, a three-and-a-half-day training exercise on Bicton Common involving fieldwork skills, obstacle crossing undertaken in Week 7 and the Baptist Run, a two-day exercise to test stalking and map-reading skills that takes place on Colaton Raleigh Common (RV19) during Week 15 of the training programme.

Officers training recruits recall when they were themselves recruits undertaking the same exercises in different areas of Woodbury Common. Crucially it is being in and re-visiting these places that powerfully brings forth these memories: 'I sometimes walk around and I can remember training here, doing this. I can remember a wood block just over the other side [of the valley] there that I harboured up in exactly. I can still think of the tree I hung my poncho up on. I could point it out now' (Sergeant Tim). Some trainers remarked that they were saddened by the manner in which the Black Hill quarry had encroached on this landscape and swallowed up areas that they themselves remembered and had trained in.

Conclusions: in and out of landscape

The relationship between the RM and the landscape is both peculiarly intense and peculiarly multifaceted. On the one hand, in order to endure aspects of their training, such as night navigation exercises following a bearing or running around the endurance course, they must learn to inure themselves from the landscape, block it out and ignore everything that it throws at them as much as possible. In this respect their experience of the landscape necessarily has to become disembodied. It ideally becomes little more than an irrelevant matrix of obstacles to the RM body getting from A to B in a limited amount of time.

On the other hand this landscape becomes embodied and remembered in an extraordinarily profound way. It is in and part of every RM body. Appropriate strategies for movement across it involve an extraordinary attention to topographic detail and the nature of the vegetation. The RM body ideally dissolves itself into this landscape, becomes part and parcel of it, all the time acutely aware of the sensory impacts of presence and movement involving touch, sight and sound. Marines must learn how to feel, observe and listen to this landscape in an acute manner. Such is their involvement in the landscape that it forms a fundamental formative part of their collective identity as Royal Marines, and significantly constructs and structures their personal biographies.

The insights from Mauss and Foucault with regard to bodily practices, routines and discipline have provided many helpful insights for this account but what is lacking in the work of both is the observation of actual bodies (persons) and routines. In this sense their accounts of the body and embodiment are simply decontextualized ideal types and theoretical constructs. Here we have attempted to construct an alternative account through field observation in a particular landscape context to produce hopefully a more nuanced and contextualized account. In particular we have attempted to demonstrate the manner in which persons and their bodies cannot be understood apart from the landscapes of which they are a part, reciprocally involved in forms of movement, action, awareness and social memory. For the RM Woodbury Common is no ordinary landscape, for what they learn there profoundly contributes to their understanding and relationship with all the other landscapes in which they find themselves. They take Woodbury Common with them throughout the world and it provides an essential medium for their understanding and recursive relationship to that world.

4
Environmentalists: the giving and the taking away

A number of anthropologists and other social scientists provide an overview of anthropological engagement with environmentalism, starting from earlier conceptions of ecological anthropology in the 1960s and 1970s, with its functionalism and systems theory approach (Kottak 1999: 23) in which an interaction between ecosystems and localized adaptations could be regarded as a functional entity (Brosius 1999: 278). Ian Scoones, citing the works of Sahlins on Fiji (1957) and Geertz on Indonesia (1963), provides examples of such thinking (Scoones 1999: 484). Debate was often informed by the then-perceived essentialist dichotomies, such as those between nature and culture, idealism and materialism, mind and body (Biersack 1999: 2, 4, 8; Little 1999: 254). Paul Little argues for the importance of environmental anthropological research that overcomes this division (Little 1999: 257), such as the work of Tim Ingold (1993; 2000), transcending these Cartesian dualisms. When coming to present-day environmental anthropology, insights are drawn from diverse sources including political ecology, economy and agency; social sources of environmentalism and local–global articulations; and the links between biological and cultural diversity. From this work arise significant issues including power and inequality, knowledge production arising from locally produced meanings, and how anthropologists represent environmental movements (Brosius 1999:278; Kirsch 2002: 175; Low and Merry 2010; Rival 2011). The issues of land use, access and management in conjunction with conservation of biodiversity are of immense importance to the Pebblebeds of East Devon, as we have shown in the previous chapter. They are also of continuing international concern and anthropological discussion.

Conflict

Christopher Timura states that 'environmental conflict' has become a master narrative in environmentalism, embodying our expectations, and when applied uncritically may actually impair analysis, mediation and policy formation (Timura 2001: 104). Anthropologists acknowledge the complexities ingrained in environmental conflict, and Kay Milton's summary is a useful starting point with which to provide a brief overview of such complexities:

> There are underlying disagreements over how problems are defined, their degree of seriousness, who is responsible for solving them, and how amenable they are to solution. These disagreements run deep; they are based on different moral principles, different values, different assumptions about how the world operates, and they are found not only at the international level, where cultural diversity is to be expected, but at all levels, within a single society or organization, and within the actions and policies of a single corporate group.
>
> (Milton 1991: 4)

Conflict comes in many guises, including within theoretical analyses. One example is that between cultural critique and activist research; the former, desirous of analytical complexity and sophistication, perceives activist research as being overly simple and unproblematized (Hale 2006: 101). However, in the detailed works of several activist anthropologists we find that it is empirically grounded and of great value. These include studies by Peter Armitage, who has worked on several projects related to land use, occupancy and the effects of both hydroelectric development and the proposed trans-Labrador highway on Innu land use (Armitage and Stopp 2003, Armitage 2006, 2007, 2010, 2011); Marcus Colchester, whose advocacy work has focused on securing indigenous people's rights to land and livelihood through policy change from grassroots to international level (Colchester 2003 [1994]; 2004); and Shannon Speed (2007), who has worked in Chiapas, Mexico, as an activist for human rights and indigenous rights since 1996. Another debate concerns whether conservation is a western concept that has been imposed on 'the rest' – but Laura Rival states emphatically that this is no longer the case: ' ... for the rest actively shapes the world's future directions' (Rival 2011: 18).

Shape-shifting

Protected Areas (PAs) have been described as a form of virtualism – they have become the device with which many people view, comprehend and experience areas of landscape: 'This virtualizing vision [Carrier and Miller 1998], … (is) a way of seeing and being in the world that is now seen as just, moral, and right. In effect, protected areas are the material and discursive means by which conservation and development discourses, practices, and institutions remake the world' [Brosius 1999, Watts 1993; West *et al.* 2006: 255–6). One darker aspect of this new cosmology is conservation of biodiversity leading to the evictions of several million people together with denial of access to their former lands (Agrawal and Redford 2009: 4; Ozinga 2003; Pearce 2005; Homewood *et al.* 2009: 5; Patinkin 2013; Brockington and Igoe 2006: 431, 454). The literature regarding the management of PAs reflects a number of ongoing concerns, including the links between community management, livelihood, poverty alleviation and development (Agrawal and Gibson 1999; Bray 2007; de Koning *et al.* 2011; McNab and Ramos 2007; Mbile *et al.* 2005; Nielsen and Treue 2012; Nunan 2015).

Further problems may be found across a vast spectrum of issues and practices. For example, forests are a symbol of people's relations with nature, providing many indigenous people with habitat and livelihood, but they are often seen as an asset to be liquidated for a multitude of purposes including timber, plantations, agriculture, and ranching (Bass 2001: 3; Brosius 1997; Hoelle 2012: 70; Colchester and Chao 2011: 8).

Other concerns include the appropriation and clearance of land by governments and multinational corporations for financial gain (Li 2010; Colchester *et al.* 2006; McAndrew and Il 2004); debate as to whether there is population growth at the edge of protected areas in countries in Africa and Latin America and what effect this may cause (Joppa *et al.* 2009, Wittemyer *et al.* 2008); and protests by indigenous communities residing in PAs or displaced by corporations' industrial practices (Kirsch 2007; Obi 1997; Turner 1995; Survival International 2013; Horowitz 2010; Evans 2007: 43; McKenney *et al.* 2004: 71–2).

It is also interesting that in his fieldwork, Jedrzej Frynas found that the most scathing criticism of protected area management came not from NGO activists as he had expected but from ' … former and current oil company staff and company consultants with first-hand experience of CSR practice in the oil and gas sector' (Frynas 2005: 581–2). The critical cultural geographer Leah Horowitz believes ' … sustainability and mining are not necessarily antithetical. There is a solid business case for good management of environmental and social issues, which often makes for a more efficient operation while reducing the risk of legal action and helping companies to prepare for future regulatory changes' (Horowitz 2006: 307). Anthropologist Rebecca Hardin adopts Marcel Mauss' notions of a social reciprocity that entails new models of mutual responsibility to analyse the environmental sector, and argues for new forms of governance at community levels (Hardin 2011: 47–8).

Management, volunteers and environmentalism on the heathlands

In this section we look first at the literature concerning volunteer motivation before providing a brief outline of other issues regarding management of PAs.

Heathland volunteers' motivations fall into five areas: having the time to take part, a strong interest in wildlife and the heathland, a desire to learn more about ecology and biology, a wish for positive social contacts, and a desire to help the environment. The academic literature confirms that these five are typical of environmental volunteers, but identifies additional ones: satisfaction in the work completed, gratitude of the people they assist and developing a pattern or structure to one's life. (Donald 1997: 489, 495; Hines *et al.* 1987: 6; Yeung 2004: 34–6, 43; Bramston *et al.* 2011: 785).

One perspective is that environmental values originate from connectivity with nature (CWN). Dutcher and colleagues state that in order for environmental management to succeed it is important to understand these values (Dutcher *et al.* 2007: 474). They write: 'Connectivity with nature focuses not on the material, biophysical, and economic interactions between people and nature (*gesellschaft*) but rather on the subjective experience that people and nature are of the same type (*gemeinschaft*)' (Dutcher *et al.* 2007: 489). This concept has been critiqued by Tam, who notes: 'Although CWN was found to be predictive of environmental concern and ecological behavior (Dutcher *et al.* 2007), the reliability of its scale was relatively low ($\alpha = .72$)' (Tam 2013: 66). Gosling and colleagues explored a similar concept – connectedness with nature – and found that there appeared to be a link between conservation protection behaviour and connectedness with nature (Gosling *et al.* 2010: 302).

The literature regarding the management of PAs reflects a number of ongoing concerns, including the links between community management, livelihood, poverty alleviation and development (Agrawal and Gibson 1999; Bray 2007; Koning *et al.* 2011; McNab and Ramos 2007; Mbile *et al.* 2005; Nielsen and Treue 2012); management of lands around the perimeter of PAs (Defries *et al.* 2007; Pence *et al.* 2003); and the identification, planning and design of a PA and its management strategies (Brandon *et al.* 2005; Kremen *et al.* 1999). Of most relevance to the Pebblebed heathland project is the subject of who does the managing and whether it is effective. Several authors remark on the change from top–down coercive conservation strategies that had led to the displacement of indigenous peoples and local communities, as well as increasing poverty and the failure of conservation projects, to an increasing acceptance of the role of community in conservation management, which is often referred to as co-management or community-based conservation (Agrawal and Gibson 1999; Basurto 2013; Colchester 2004; Reid *et al.* 2004; Torkar and McGregor 2012). This includes acknowledgement of the importance of local participation by the International Union for Conservation of Nature (IUCN): 'the recognition of indigenous and community conserved areas (ICCAs) that fully meet protected area definitions and standards in national and regional protected area strategies is one of the most important contemporary developments in conservation' (Dudley 2008: 29).

Krueger explains that good governance is dependent on a number of factors, including transparency and accountability (Krueger 2009: 24). A global study into the effectiveness of PA management has been conducted to help the conservation community share best practice. Adequate equipment and infrastructure were found to be vital in good management practice but the majority of regions had only inadequate provision (Leverington *et al.* 2010: 47).

The politics of environmentalism

Environmental and conservation practices and the rights of indigenous populations are often inextricably linked to global capitalism and the rhetoric and practices of many multinational corporations (Pendleton *et al.* 2004: 58; Ruggie 2008: 14, 51; Coumans 2010: 46; Conley and Williams 2008: 34; Welker 2009: 166). Ruggie describes the root cause of the human rights and business predicament as lying in the 'governance gaps created by globalization' (Ruggie 2008: 3).

The literature concerning conservation practice shows that knowledge, organization, accountability, and social justice are all important factors in the arena of conservation of biodiversity. Brechin *et al.* (2002)

state: 'Consensus on the question of who the conservation community ultimately serves and how will define the degree of legitimacy that the biodiversity protection imperative will take on for those resource-dependent populations whose livelihoods and oftentimes survival depend upon nature's vitality' (2002: 58).

Environmentalists and the Pebblebed heathlands

Approximately 80 per cent of the world's lowland heathland has been lost, and 25 per cent of that remaining is found in the UK. Maintenance of the Pebblebed heathland is therefore a huge responsibility. This section focuses on the vital contribution made by those who volunteer their time, energy and skills in helping maintain and contribute towards the Pebblebed's recovery as a heathland landscape.

Heathland wildlife

Before she started volunteering with the RSPB and understood what the heathland is all about, one volunteer said that she had thought it a 'bit scruffy, dry and desert-like'. To an uninformed observer, this landscape may appear barren of much life, but this is far from true. On the Pebblebed heathland, of the bird species that are unique to lowland heath in the UK, the foremost would be the Dartford warbler, which a 2011 survey showed had dropped in numbers to about twenty pairs (it is thought that the summer weather conditions and previous winter's snow were the reasons for this). Other bird species that tend to depend to a greater or lesser extent on the lowland heaths include linnet, meadow pipits, nightjars, stonechats, tree pipits and yellowhammers. Snipe breed here, with large numbers of them over-wintering on the heathland too. The woodland fringes are also very important and are home to the bullfinch, song thrush and the mistle thrush as well as a few other species. Other breeding species include curlews and woodcock; recently longhammer have started breeding here, which an environmentalist describes as being a first for Devon in at least thirty years. Buzzards and kestrels are often to be seen sweeping the skies over this landscape and are part of the seventy-plus species of bird recorded here by the RSPB.

An important feature of the heathland is its insect and plant coverage; on the RSPB reserve (and in other areas of the heath) dwell the greatest number of butterfly species recorded in any RSPB reserve.

Thirty-eight species of butterfly having been recorded since 1997 along with twenty-three species of dragonfly and damselfly, of which eighteen are known to breed here. The Pebblebed heath is home to the silver-studded blue butterfly (*Plebejus argus*) and the southern damselfly, both of which are endangered protected species. Of course there are beetles, including Kugelann's ground beetle, which is very rare and not found in many places at all, and ants, and it will be seen just how important one species of the latter creature is to the silver-studded blue butterflies. Amphibians such as the great crested newt, which due to its decline in numbers is a protected species, frogs and toads dwell near the pools and ponds; reptiles are to be found too, including adders, grass snakes, slow-worms, smooth snakes and at least two types of lizard.

The heathland soil is nutrient-poor, acidic and home to coarse grasses and over 450 species of plants, including the rare allseed with its incredibly tiny leaves, the bee orchid, so named because of its bee-like structure, bog asphodel, common butterwort, cross-leaf vetch, grassleaf vetch and two types of gorse (one of which, dwarf gorse, the Dartford warbler likes to nest in and another, the taller European gorse, on which it likes to perch to look around its heathland habitat). There is also heath milkwort, heath violet, three species of heather (bell, cross-leaved and ling), pale butterwort and the insectivorous sundew. Also to be found is dodder, an interesting parasitic plant that may be found in deserts but here lives on the gorse, upon which it depends for its chlorophyll.

Volunteer groups

The majority of those interviewed take part in the weekly Wednesday RSPB volunteer group but there is also a regular monthly Sunday work party too, and others we spoke to had specific interests in safeguarding and promoting the habitat not just of birds but also of butterflies and damselflies, particularly those which are endangered species.

The volunteers we spoke with, aged between thirty-six and eighty years old, come from diverse backgrounds and include those currently or formerly involved in business administration, chartered surveying, farming, histology, management in education, running a smallholding, social work, tattooing, teaching, tool making, transcribing and warehouse management. All share a common interest in wildlife and bring their invaluable skills with them, including IT literacy, which has helped with the entry of records, and carpentry and tool repair, which has ensured tidiness and efficiency at the barn in which the RSPB office is based.

Why volunteer?

There are, of course, individual reasons for volunteering. One is found among those who have recently taken a degree in subjects such as conservation, environmental protection, or ecology, and who wish to become employed in their chosen area. Without training and hands-on experience of working in conservation, familiarity with various habitats and tools, and experience of outreach work and engaging with the general public, the possibility of getting a job in the conservation industry is remote.

> I wanted to get my chainsaw license, not just the badge, but to be working on it all the time so I could become a competent user and get experience of using the tractor as I've already got my license for that and I just like the heavy machinery that they've got here – winches and chippers and mowers and things like that. I also get to do the heathland management and working with the team up there; it's just brilliant.
>
> (Conservation intern)

This intern goes on to explain how his chainsaw licence has been paid on his behalf and that this has saved him at least £800. For others, after long-term interest in wildlife generally or birds in particular, volunteering with the RSPB seems a logical move: 'I retired and I've been in the RSPB for, well, thirty-five years and I've always promised myself I'd do some volunteering when I got some space in my life I suppose, so, that's what I've done.' Volunteering is thus a way of putting something back into the RSPB. In fact, several of those interviewed have retired and one volunteer provides a particularly interesting and considered account of what volunteering means to him. He has taken early retirement and describes how his friends say his life must now be one big holiday, but for him this is not the case:

> There's things that I do that some days I don't really want to get up and do. It's a nuisance but I've committed to doing it. You've got to have commitments and some meaning in life. If it's all one big holiday ... well I'd go back to work after a while because you just can't, you've got to have a reason for getting up and, there's probably a feeling that you're actually putting something, doing something, achieving something. I choose how much I can do and when I can do it ... I don't do it to say 'Oh look, I'm doing this, aren't I good,'

or, you know, a lot of organizations who as soon as people get a job they're given a badge or a chain to wear round their neck or whatever. I'm not interested, you don't do it for that, you do it for a sense of achievement or just having some meaning in life. And there's a pattern to it, the fact that you go every Wednesday, in addition to you knowing which day of the week it is one day a week, 'Oh there's another week gone by,' so that's what it provides, but not that, not to be able to go around saying, 'Look what I do; I'm wonderful, I'm doing this for the community,' because I'm being selfish really, I'm just doing it for myself.

(Retiree volunteer)

Having the time to become involved in volunteering is often a reason that is given: 'I volunteered in 2000 because there was a nationwide campaign called 'Time Bank' and they were just advertising if you can't give money then you can always give time. I just love being outdoors, really absolutely love being outdoors and I enjoy gardening and I enjoy nature and learning' (Conservation volunteer). Another volunteer describes how once her children started school she found she had some time on her hands and became involved with butterfly conservation:

At that particular time they wanted people to record silver-studded blues and of course I was living very close to this area, which was ideal for silver-studded blues and a friend and I found a couple of colonies on the Commons and then we just kept on recording them and then the records got wider and wider and it all developed from there really.

(Conservation volunteer)

Other volunteers have come from backgrounds where they have done voluntary work in other spheres, such as helping at a hospice or with organizations such as the National Trust where much of the work they did was outdoors. We look next at what volunteering entails.

Giving and taking away

In the early Bronze Age humans started to open up what was probably previously a heavily wooded area, cutting down the trees for firewood and preventing them from re-colonizing by grazing animals, which led to trampled bare ground and areas of high and low vegetation. That continued throughout history. Once these spaces were opened they were kept open by human activity. The heathland is, therefore, known as a plagioclimax, a human-made environment. In former times the heath was actively exploited by local people, so remaining an open landscape. The trees and gorse were a source of fuel, the heather a cheap form of thatching and the bracken was used as animal bedding. The heath was constantly regenerated by the clearing and coppicing of such materials. It would then be swaled, and this would result in heathland regeneration and a varying vegetational age structure throughout the heath, something now referred to as a 'mosaic', which is of great importance to the heath's role in providing the appropriate habitat for its wildlife. This then was a complex form of maintenance and emulating it is now the task of environmentalists in the twenty-first century.

Starting work

At the beginning of a volunteer conservation session the volunteers are informed of the tasks to be completed that day. The experienced majority is used to a routine and understands why they are working on a particular task and so a full briefing session is not required. However, when it is recognized that the knowledge is not implicit, for example if there is a new volunteer or someone who has not experienced working with certain tools, then the team leader may either give a full team briefing or will explain to the individual what is to be done, why it requires doing and what tools are necessary for the project. The feelings of the new volunteer are well understood: 'It's difficult at first until you've got used to the routine but you learn a lot from talking with the others and if you're just given a lecture about something in one hit, you don't take it all in anyway' (Conservation volunteer). Once the work commences a new volunteer can learn a lot from observing how others use the tools and the team members are always pleased to answer any questions a colleague may have: 'It's something (knowledge) that accumulates but if you don't know, you ask. The atmosphere is such that you can say, "well, why are we doing this?"' (Conservation volunteer).

Tasks and tools

The tasks are connected to the time of year and what is happening in the landscape. Breeding and nesting birds must not be disturbed and so it is in winter that the birch and willow are cut, the logs stacked and the bracken burnt. Paths are cleared, often using scythes to cut back anything encroaching on the paths; road ruts are filled, bridges erected over wet boggy land and boxes created for dormice and bats: one volunteer reported 'My bat box was like the Leaning Tower of Pisa [she laughs] but we also do some other interesting one-off things like felling the tops of the birch trees that were then made into faggots for use on steeple chases' (Conservation volunteer). This volunteer describes how the birch are felled with a billhook, which she describes as like a little machete, broad with a hook and 'quite lethal'. In fact quite a large variety of tools are used by the Wednesday volunteer team – various saws including a bush saw, loppers, slashers, spades for digging holes, a bracken bruiser, tampers for tamping in poles, croppers for cutting silver birch, hammers, axes, ropes, chisels: 'We probably use most things other than items that require proper certification, in terms of chainsaws, I mean, there's no way you'd even contemplate thinking, "Oh, I'll just have a go with that"' (Conservation volunteer). Whatever job is being done it is important to ensure the correct tool is being used and that the tool itself has been properly maintained: 'If it's not sharp or up to its function then it makes life harder' (Conservation volunteer).

It is also vital that the volunteers look out for each other:

> Yep, I think a lot of what we do is potentially dangerous but I think because of the age group and the background that people have come from, it's an unspoken issue. You are automatically watching out for each other and we keep an eye out on where the tools are; if you put a fork down, you put it in the ground; so it's all done sub-consciously, really.
>
> (Conservation volunteer)

This volunteer went on to remark that when slashing bracken he goes into a rhythm, and finds it therapeutic because he gets in to the swing: 'Well, people have either got some natural, well it can't be said 'talent' for slashing [both volunteers roar with laughter], talent, or whatever but they've got some common sense.' He believes it is essential to possess a degree of practical sense and feels that some people do not stay with the group for very long because they do not think about how to do the job, or what tools to use and how to use them: 'So there's a way of thinking but how consciously you think about it, I don't know.'

One volunteer is aware that uninformed people may be alarmed at the cutting down of trees: 'People are so programmed to believe that trees are a good thing, "Why are you cutting them down? Why are you making fires?" and this sort of thing, so often a little curtain of trees is left to separate us and our activities from the public.' However, signs are now put up explaining why tree clearance and cattle grazing are taking place and this volunteer feels that this helps people to accept what has to be done:

> I guess I was one of those when I said I found the Commons kind of ugly at first (several years ago). You know, you see burnt bits and you see scraped bits and it's not very pretty and you don't understand why it's done … this is where explaining is good.
> (Conservation volunteer)

This volunteer points out some of the benefits of disturbing nature, for example, where the cow's hoof imprint collects the rain, the damselfly then lays her eggs in the imprint. It is clear that conservation and education go hand in hand.

Two other environmentalists comment on how the mosaic on the heathland is created. One says, 'Everything we do to manage it uses fossil fuels, uses petrol or oil, and it's such a huge influence on this area. I think it's just too much, too intense what we're doing' (Environmentalist volunteer). This volunteer feels that the result of this is too many plants of a single age structure and that the work areas covered should be smaller in dimension, which might help alleviate this problem. The second, speaking on this subject, also comments on how using heavy machinery tends to leave the habitat uniform, and feels that more cattle-grazing in particular might be better for the wildlife. However, both acknowledge that with such a large area to cover and only two wardens, plus the RSPB and volunteers working in this landscape, creating a mosaic more similar to that which obtained when the heathland was used by commoners is virtually impossible.

Swaling, an art form in itself, is also undertaken during the late winter. One interviewee informs us that many years ago the villagers of Woodbury would go up en masse to the common and do the swaling. More recently she took part on one occasion. She laughs, saying they failed, probably because there had been a frost; it was very cold and the fire would not catch.

In the summer, when the cattle go out, electric fencing is erected for their safety and the fencing is regularly inspected to make sure nothing is touching the wires (otherwise the electric current would not conduct but divert down to the ground). Bracken and brambles are cleared: 'I've got the scars on the backs of my legs to prove that' (Volunteer). The wardens or qualified interns mow the firebreaks using special tools fixed to the tractor. This clears the ground of gorse, bracken and small trees to enable easier control should a fire start. The volunteers rake the material in to piles, which they then fork in to the hopper (a container that tapers downward and is able to later discharge the contents): 'We don't want the nourishment to stay there for the plants that grow on the heathland grow best on impoverished soil' (Volunteer). This is hard work, one volunteer describing it as backbreaking. New skills are learnt and two volunteers spoke with pleasure about building fences, gates, lean-tos and barns: 'You learn from working with other people and two or three of us are now expert barn builders whereas we weren't when we started' (Volunteer).

Also, in the summer, different types of survey are conducted. These cover butterflies, dragon and damselflies, dormice, Dartford warblers, nightjars, heather, gorse and other types of vegetation.

Butterflies

Among the thirty-six or more species of butterfly that have been recorded on the heathland, the silver-studded blue is an endangered species nationally. It is only the male that is blue, the female is brown and it is thought this gives her better camouflage, although in flight her wings too may look somewhat blue. The species is now extinct in Scotland and the north of England; surviving populations are mainly found on the Hampshire and Dorset heathlands so its presence on the Pebblebed heath in East Devon is greatly welcomed, and may be an example of what has been described by the charity Butterfly Conservation as 'an indicator of active lowland heathland management' (Ravenscroft and Warren 1996: 4). The silver-studded blue shares a symbiotic relationship with the common black ants, *Lasius niger* and *Lasius alienus*. The ants are attracted to the sugar-rich liquid produced by the larvae and pupa and carry them off in to their nests, where the ants tend them from hatching and through pupation. In return for the ants' protection from parasites and predators, the larvae and pupa allow them to gather this sweet liquid from their nectary organ. When the adult butterfly emerges from the ants' nest, it is often covered in ants gathering the last droplets from its surfaces. As well as their ant protectors these butterflies require open ground on which to breed and bare soil or short vegetation. Not only do these conditions provide a warmer microclimate at ground level for the larvae, the larvae themselves exist by eating the tips of young heather.

The butterfly survey is conducted throughout the flight season, which is usually six to eight weeks from mid-June, though the volunteer coordinator starts looking out for them from early June as they sometimes start flying early: 'It's a lovely area and I look forward to coming, to re-establishing my contact with my butterflies' (Butterfly conservation volunteer coordinator). Together with six other volunteers who help her she makes a record once every ten days; the butterfly's life span is a week so she knows that she will be getting different butterflies on each count. The locations where the silver-studded blue may be found are divided into nine or ten areas and methodically counted. The numbers vary in each area – when the vegetation becomes too tall, usually after a couple of years, the butterflies move elsewhere and it usually takes three or four years after the land has been cleared before the butterflies arrive, as the ants and heather have to establish themselves first. Sandy soil for the ants, bare patches, young shoots and flowering heather of the ling (*Erica cinera*) variety are what the butterfly conservationists of the silver-studded blue are looking for on the Pebblebed heathland.

During the course of the interview, held whilst we conduct a survey, we also recorded several other species of butterfly, moth, hoverfly (frequently mistaken for a wasp because of its yellow and black colouring) and bee; these include green hairstreak, small pearl-bordered fritillary and small heath butterflies, and common heath and grass moths. The coordinator views ground-scraping as the best method of clearing the land for the silver-studded blue and this has brought her into conflict with archaeologists, who find that topsoil scraping damages or even destroys archaeological evidence that can never be replaced (see Chapter 2).

Damselflies

On Aylesbeare and Colaton Raleigh Commons dwells the endangered southern damselfly (*Coenagrion mercuriale*), a tiny creature less than three centimetres long, which takes its scientific name from the shape of the winged helmet of Mercury on the male's abdomen. As much as twenty-five per cent of the world's population may be found in the UK, and this is a species that is viewed as being in danger of extinction internationally (Devon BAP 2009). Management of grazing by cattle is important to control black bog rush, western gorse and willow scrub on the vegetated runnels and streams in these locations. Surveys have shown an increase in the dragonflies' numbers at Aylesbeare, probably because of improvements in the management of cattle in particular but surveys on Colaton Raleigh indicate a decline and it is believed this may be due to the effects of insufficient grazing in this location. Sadly the southern damselfly no longer dwells at Venn Ottery and was last seen there in 1990. The volunteer conducting the surveys has been doing so for over twenty years; she and her husband introduced grazing with Ruby Red cattle on Aylesbeare and Venn Ottery to help manage the sites for these damselflies:

> I'd go up there every day and check them and stuff and when the girls were little it tied in nicely because I'd take them with me but we've moved on to different things now. However, all the places we used to graze are still being grazed by other people and in fact the Clinton Devon Estate have bought our three cows and ten other youngsters and they're grazing and they're going to hopefully continue to do so with their own animals, which will be ideal.
>
> (Butterfly conservation volunteer)

The Ruby Red is the perfect cow to graze on the heathland. It comes from Devon, has a bulky body and short legs and quickly puts on fat, which means it does not feel the cold as much as other cattle; importantly its huge appetite means that it is willing to eat black bog rush, which helps secure the habitat for the southern damselfly.

Nightjars

During Heath Week there are organized trips to see and hear the nightjar, which is a popular bird amongst visitors to the heath. They are nocturnal and crepuscular and hawk for food or look for mates to breed with at dusk and dawn. With a near-silent flight and loud churring call that may be heard for up to two kilometres on a still night this bird is often more heard than seen: its plumage is grey-brown, streaked and mottled, providing ideal camouflage during the daytime when it hides in the low vegetation of the heathland. The species has fascinated humans over the centuries. Many years ago this bird was sometimes known as the 'goat suckler', as at night they would sit on goats' warm bodies to feed and people believed they were suckling milk. With their low weight-to-wing ratio they fly low and slowly, capturing moths and beetles with a mouth that contains hair in which to lodge their food.

One of the volunteers has taken part in a nightjar survey around Woodbury Castle. She explains that it was quite difficult as she had to cover both sides of the road and as there is only a small window of time in which to hear the birds churring, from sunset and for the following half-hour. She made several trips to walk the location and listen out for their special call. She heard quite a few nightjars and was also surprised to find how busy the Common is at night:

> When I was up there I didn't expect to see any people at all but there were Marines, there were bird people, people interested in hearing the nightjars, people parking, I don't know what they were doing (and in a quiet tone of voice) – yes, *that's* what they were doing [interviewee and interviewer both laugh].
>
> (Conservation volunteer)

Another volunteer who has taken part in Dartford warbler surveys describes how volunteers listen to the birds' song, determine whether it is a male or female warbler and record the number in each location. Three of the volunteers have taken part in surveying the heather and gorse. Each of them takes an area of the RSPB reserve over a five-year period and each year covers two or three patches in 50 m square grids. The grid point is visited, photographed and the flora and vegetation recorded.

RSPB volunteers also often keep their own records of what they see on the heathland both when working and when out walking in their own time. This information is passed on and the data entered on the RSPB computer. One of the volunteers has taken responsibility for recording such data and, as well as participating in the Wednesday Group, regularly visits the RSPB office at other times to ensure the records are kept up to date. He believes that this encourages people to keep records:

> If they can see that you're actually taking that information and putting it in, then they're more likely to record whereas if people think, 'Oh I make these records but then what happens to it?' they won't bother.
>
> (Conservation volunteer)

There are also other individuals who are not part of the RSPB volunteer team but who regularly come to the heathland to look at the wildlife, including the Dartford warblers for example; they too keep their own records but it is not known just how much of this data is recorded. One person we spoke with has a great interest in spiders and keeps his own record book of what he finds when he visits with his wife:

> I was quite surprised, from the natural history point of view, the birds, the butterflies and the botany, the three Bs are always well covered. When you get down to the invertebrates and beetles, they are usually well-done as well, but spiders and things like that are just ignored; nobody does anything about it but yet you've got these superb heathlands there and so I began to take interest in that.
>
> (Heathland visitor)

This visitor would like a proper survey to be conducted, perhaps by the Devon Wildlife Trust, and describes how different spiders live in the different habitats on the heathland, some favouring the wet heathland and some the dry: 'In the wet heathland you get a group called *Pirata*; the common one is *Pirata latitans* but you also find *Pirata piscatorius*'. The online National Recording Scheme for spiders now regards the *Pirata piscatorius* as being in decline in the UK due to the degradation and destruction of many bogs and wetland areas associated with standing water. Yet again, the Pebblebed heathland is providing a habitat for a species that may be endangered. Not at risk but of some interest is the *Agelena labyrinthica*, the funnel-web spider, whose lacy creations can be seen spread among the low vegetation:

> When creating her web the female builds its tunnel to secrete back in to the gorse and towards the end of the season she makes an egg cocoon and then builds an intricate door around it with lots of little tunnels in it. The theory is that predators get in to here to various tunnels and they wander around and they can't find the egg cocoon, which is right in the middle somewhere, hence labyrinth, so that protects the eggs and then the young spiders over the winter.
>
> (Conservation volunteer)

The RSPB volunteers work in many areas of the Pebblebed heathland, not just the RSPB reserve at Aylesbeare but also Blackhill Quarry, helping its return to heathland by clearing birch and pine, and at Colaton

Raleigh, East Budleigh and Venn Ottery commons. The RSPB often works in partnership with the Devon Wildlife Trust, helping with grazing animals and preparing the ground for the importation of species such as damselflies. As well as helping out Clinton Devon Estates, on occasion they also move off the heath and assist other organizations, for example East Devon District Council, Sidmouth Town Council, Blackdown Hills Area of Outstanding Natural Beauty Partnership; the RSPB also helps to restore quarries for Aggregate Industries in the local area.

Feelings about volunteering

Several of the volunteers mention how they visit parts of the heathland that they had not been to prior to becoming a volunteer. Working with the RSPB has extended their knowledge and understanding of the heathland: 'Slowly you're amassing knowledge and you're remembering that's what we did three years ago and that's what it now looks like now so there's satisfaction in seeing what you are doing' (RSPB volunteer). Another remarks: 'You get to bits you probably wouldn't have seen and you're helping to bring it all together, but even if I didn't belong to the RSPB, I'd be going up there walking, it's just that my knowledge, the depth wouldn't be the same. Give it another ten years and it'll be even greater!' (RSPB volunteer).

The experience of working as part of this team is enjoyed and found to be satisfying, and not just in the work completed: 'There's satisfaction and that satisfaction doesn't necessarily mean that you stand at the end of the day and say, "Oh, we've cleared that quantity of thistle or scrub"; there can be satisfaction in the way you've worked together as a team or with somebody' (RSPB volunteer). The work is often physically challenging and one volunteer remarks:

> I think that often we work harder as a volunteer than you would if you were being paid; particularly when on some of the jobs that I've done, there's been a lot of stuff to be moved, it's very exhausting. I go home staggering sometimes but very happy, and there's great camaraderie.
>
> (RSPB volunteer)

There is great commitment to attending the Wednesday session, whatever the weather, some of the volunteers commenting that new people either tend to come and go quite quickly or become very committed: 'It's quite remarkable really, I think, the way that people embrace it the way they do ... once people commit, they seem to, you know, they even come in, sometimes apologetic, that they've missed the last two weeks' (RSPB volunteer). This commitment is exemplified within the words of one of those we spoke with, a former toolmaker; he describes how he uses his skills at the barn: 'I go on a Tuesday and I do a lot of work over there, keeping the tools in good order but sometimes making tools, benches and shelves' (RSPB volunteer). He prefers working in the barn on a Tuesday, when he can use his skills and go at his own pace, to the work on a Wednesday:

> I'm now at a stage where I'm getting trouble with my legs and I'm finding it more and more difficult because we do a lot of walking of course on a Wednesday. Last Wednesday I stopped at four, I'd had enough, and it was about eight o' clock before I could bring myself around to get myself any food.
> (RSPB volunteer)

Many reflect on the sense of 'belonging' that comes with working in this team, with newcomers being welcomed and a definite pride in the group and the work it does. The experience is also a sociable one:

> We just seem to get on and people come from a diverse variety of backgrounds, and it's very sociable, nobody really dominates ... the social side is important; I'm sure that's why people, it's certainly why I still go and probably do as much as I do.
> (RSPB volunteer)

He remarks at how he may be getting bored with a task but turns to one side and sees an eighty-year-old still there, working hard, and thinks: 'Oh, blimming hell, I'd better keep going.' In fact, the camaraderie referred to above is motivational: 'There's a lot of banter about, "Oh, we've got to keep going"; there's a point where you do want to stop but there's also that, "Oh, we must get to that corner at least." We keep each other going' (RSPB volunteer). Several of the volunteers remark upon how they are never pressurized by the wardens to do a task or take on extra work and are always thanked at the end of the day: 'Well, it does make all that difference, doesn't it, just that word of thanks' (RSPB volunteer). When discussing a shortfall in volunteer helper numbers during Heath Week 2009, one of the volunteers states:

> They're a bit short (of help) but the trouble is, you see, they're so good, that they don't ask you. It's only if you become aware that they might do, then you sort of stick your head above the parapet and say, 'Do you need some help?' They won't come to you and say, 'Oh, can you come and help us?' but I think that's why people keep going and helping them because they don't feel as if they're being forced into doing things.
>
> (RSPB volunteer)

For several, volunteering and the heathland has become an important part of their lives: 'You take possession of it, in a way, ownership of it' (RSPB volunteer). This sense of 'ownership' is closely linked to responsibility. For example, when out walking someone may see litter and pick it up; somebody else may notice work that needs attending to and report this back to the warden. There is, in fact, affection for the group and being part of it: 'Affectionate, is how I feel. I feel that it's a local thing and I've been let in, this is how I feel' (RSPB volunteer).

Feelings about the heathland

All of the environmentalists we spoke with also spend time on the heath on occasions other than when volunteering. The majority walk, although one also cycles and runs and another visited in his childhood and rode his horse there:

> Yeah, our parents used to get frantic, 'Oh you mustn't go anywhere, they've got military exercises'. Of course that was a magnet for us; we were able to see the exercises and have a bit of a run in with the military, it was great fun, hugely (roars with laughter), it was great sport, yeah! And the ponies loved it, the horses loved it; they love a bit of battle. Even when I was going up with my own kids when we lived nearby, we'd take the ponies up there and I remember once there was a battle going on and I thought the ponies would flee but not a bit of it; their tails were up, phrow, phrow, phrow.
>
> (Conservation volunteer)

One volunteer who runs on the heathland finds it an excellent way to explore the landscape, choosing the narrow, defined tracks on the heath and the variety of tracks to be found on its edges in the woodland. He says if he finds himself at a dead end he can turn round and run back again: 'Whereas if you've walked down it you're much more, "Oh God I've got to walk back now the way that I've come."' If you run it, it's not such an issue. He does his best to avoid people, particularly dog walkers, and sometimes the pebbles themselves if they are loose ones as he finds running on them can be like running on ball bearings: 'But you don't avoid that route because of them, you just, on that path, that line, you just take what looks the easier line.' He has found that his walking experience on the commons has changed since he became a volunteer and his walking gait has become slower. He takes more time. Whereas his interest was once focused on birds he is now much more aware of butterflies, dragonflies and flowers: 'If you see something you stop and if you don't know what it is you take time to work it out.' He keeps a rough record of what he has seen and has recognition sheets to refer to. Taking photographs on a digital camera and showing them to other volunteers also assists in working out what he has seen. He feels that when he used to work in London he was 'switched off', and that volunteering has re-awakened an interest in wildlife that he had in childhood. In this sense the heathland has become an important part of his life.

Maps produced by environmental volunteers focus primarily on areas in which they have worked, the Aylesbeare and Harpford RSPB reserve in particular and Bicton and East Budleigh Commons. All of them mark important areas including bird territories, places frequented by nightjars, Dartford warblers and stone chats.

Others show observed butterfly colonies and places where lizards can be seen. One map of the Aylesbeare RSPB reserve marks ponds, experimental vegetation monitoring plots, wooded areas, a bat house in a Second World War bunker and other details such as the unwanted presence of Leylandii trees on the reserve boundary and a caravan where the retired warden used to live, as well as the line of the gas main.

These maps provide a very good indication of place preferences in relation to environmental knowledges.

For one environmentalist, his relationship with the heathland and whether he feels part of it when he is in the landscape depends partly on the weather: 'It's a deciding factor, yes, and to actually put my finger on what it is that makes it that much more special on a particular day compared to another day, I don't quite know' (Conservation volunteer). He compares himself to his wife, who can sit, contemplate and reflect on the beauty of the place, whereas he cannot 'switch off':

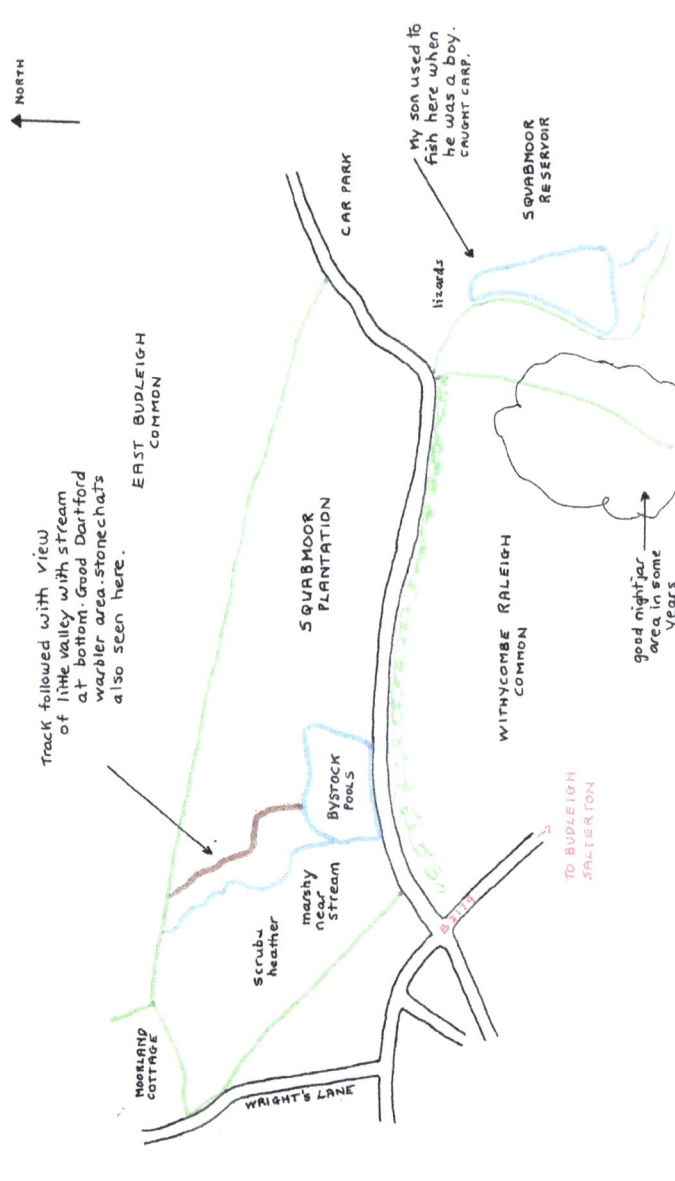

Figure 4.1 Environmental volunteer's map 1

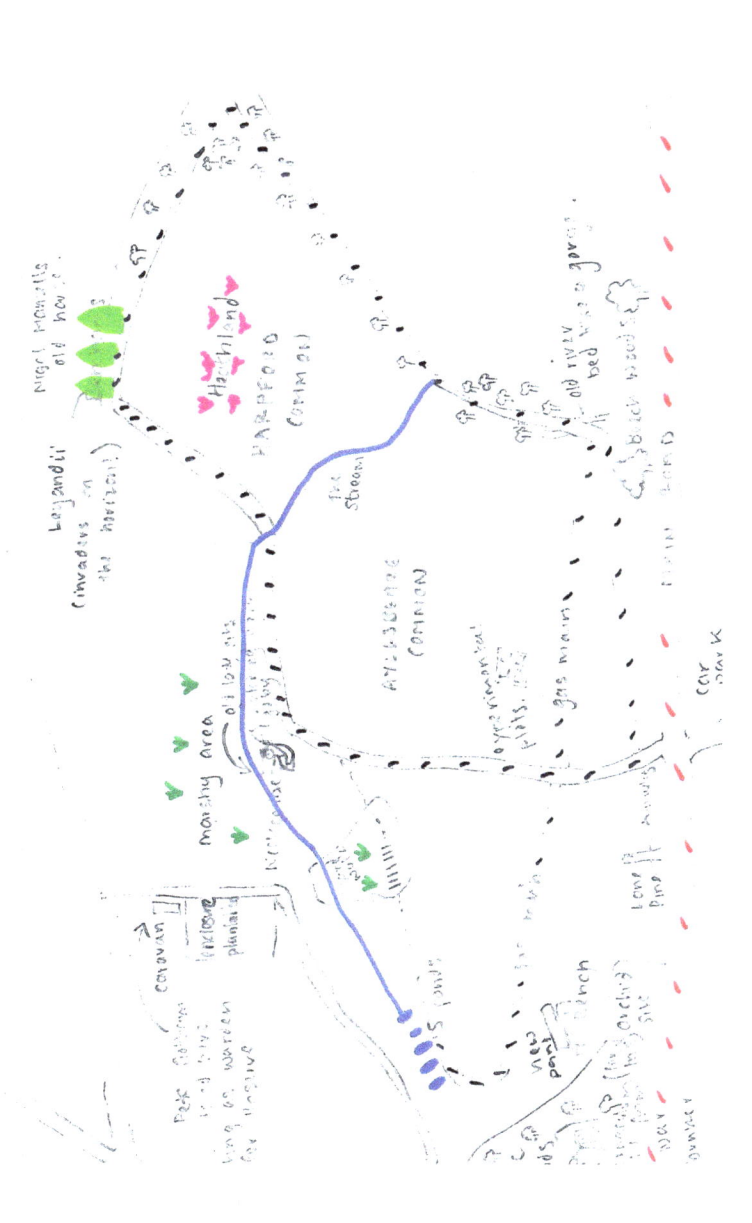

Figure 4.2 Environmental volunteer's map 2

> I can't be divorced from what's around me. Ever since I was so high I've been naming things and it continues now and I find it very difficult to switch off and not continue to name. If I can't name it I get frustrated or I have to make a note and try and find out about it.
> (Conservation volunteer)

One team member stated that being on the heathland, being in nature, is a gift and a privilege. She feels it re-builds the energies that might otherwise be lost in her day-to-day living and in this way the heathland gives her far more back than the time and energy she gives to it when volunteering. Other volunteers speak of stopping work on certain sites just to look at the beautiful views, the latter proving to be restorative; the Pebblebeds are a therapeutic landscape for many. One describes the place as an extension of her garden: 'I call it extreme gardening, with what I do up there. It is! The work I do up there gives me ownership of it, a little bit, to work like that' (Conservation volunteer). She finds that she is now confident when walking on the heath as the familiarity with this landscape that has come from volunteering, and appreciating the beauty it contains, has brought a calmness to the experience. Like all the environmentalists interviewed she does not want to see any amenities constructed on the heathland and feels that promoting it as a tourist attraction would create a dilemma; that there is a delicate balance between having the facility of nature and encouraging people to enjoy it and avoiding too much human presence: 'The public pressure can destroy what they have come to look at' (Conservation volunteer).

The RSPB volunteers are well aware that this is a human-made environment and is therefore not a 'natural' place in the purest sense, but some do feel that this requires qualifying. Some find the heathland's substrata natural; another speaks of it being natural in the sense that an opportunistic Nature has taken advantage of human interference and one speaks of how it is 'natural' for a short period of time before clearance work has to be started to prevent the progression of scrub and then woodland. 'Not manicured', and 'it has a natural feel about it' are two other viewpoints, but some volunteers do not regard it as natural at all: 'No, not at all. As a matter of fact, nowhere in England is natural. If it's there it's because somebody put it there or somebody left it there, so, no, it's totally unnatural' (Conservation volunteer). One environmentalist emphasizes that this is a plagioclimax in which humans have prevented the ecosystem from developing further but that it is not the same as modern farmland, adding: 'It's a landscape that's got all the politics from the Iron Age to now played out there; all the politics is laid out there; the whole of history is laid out. It's a wonderful history book' (Conservation volunteer).

The question regarding whether the Pebblebeds is a wild place also provokes a similar response, some believing that parts of it are wild, others finding it wild, and one speaks of it as being pristine. Like several of the walkers and cyclists, some of the environmentalists compare the heathland with other terrains they have walked in and found to be more wild – Scotland, Dartmoor, the Lake District – and one, again, qualifies his response that it is not wild by stating: 'But, at the same time, you can, if you get up earlier enough in the day, get there before the dog walkers, then there is a degree of remoteness relative to the environment you normally live in, yes. But it's not wilderness' (Conservation volunteer). Another asks:

> Have you read *Return of the Native* by Thomas Hardy? Well I think when you read that you get the impression of really wild country whereas obviously now because of the development all around it, it isn't so wild and it probably wasn't wild in those days, it was just under-populated. I think I tend to look on an area as being wild if it is unpopulated or under-populated, not visited very frequently, so in that instance I would not call the Pebblebed heathland wild. Because you can never get away, there always seems to be somebody around.
> (Conservation volunteer)

Their words used to describe this landscape indicate their feelings about it: 'magical', 'attractive', 'valuable', 'ancient', 'accessible' and 'varied' are all referred to, with one volunteer describing how she finds the landscape's rustic beauty to contain a very positive energy; that it is a place of freedom.

Conclusions

When speaking with the RSPB volunteers one senses their overwhelming commitment to the work they do on the Pebblebeds and, despite their great interest in wildlife and conservation, their awareness that the volunteering is not done for purely altruistic reasons. There exists a giving and taking in this landscape. The volunteers give of their time and, in taking away the vegetation that would lead to its destruction as heathland, important wildlife species are given the opportunity to survive, if not thrive; this, in return, provides the volunteers with immeasurable satisfaction. In this human taking away there is a giving back by the heath. This is more than just a temporary imprint on the landscape, or a temporary exchange between people and nature; much of the work done is rotational, annual in many instances, and is thus continual and creative.

Another important element is the sense of belonging that is felt by many team members coupled with a temporal sense of ownership and attachment to this landscape and its environs; several of the team speak of proudly showing their family and friends around the areas in which they have worked. The social aspect of being a team member is important and to do with the contact and exchanges between people, a knowledge that each feels the physical challenge of many of the tasks and a shared appreciation of having worked successfully as a team. All the people we spoke with told us of their desire to know more, to learn, understand, and to exchange their knowledge with their colleagues.

Although Krueger (2009) is referring to the conflict between the rights of indigenous and local people who find their access to resources is restricted by certain conservation practices in protected areas, some of the issues she raises (transparency and accountability, decent equipment, infrastructure and best practice on community participation) are also of relevance here when discussing environmental volunteers. Being involved in the conservation of nature is therapeutic, and it seems that part of this sense of well-being felt by the heathland volunteers is due to the way in which management involves them, respects and is grateful for their participation. The feeling of inclusiveness is key and should be of interest to other conservation groups involving volunteers. Overall, the feelings that the volunteers have expressed are part and parcel of the reason why they volunteer, what motivates them. Milton writes:

> Although it might seem self-evident that emotion is central to motivation, the point still needs to be argued, rather than merely stated or assumed, because the emotional character of motivation has been largely neglected in recent decades by those scholars who have sought to understand it.
>
> (Milton 2002: 92–3)

The volunteers' feelings and understanding that this heathland landscape and its diversity are valuable, and their desire to help maintain its viability and develop further knowledge about its flora and fauna, are a clear example of what Milton describes as the myth of an opposition between rational thought and emotion. She describes how rational thought is motivated thought and that emotion is the 'essence of motivation … rationality is itself a feeling, it is emotionally constituted. It is the direction provided by emotion that makes thought rational' (Milton 2002: 150).

It seems that taking part in conservation of the heathland is of great importance to many of the volunteers' lives; it has become part of them, who they are, what they do and why they do it; part of their social being. In this sense it is part of their identity, a dynamic interrelationship that encompasses material bodily movement and emotional involvement. Embodiment here is multifaceted and deep; it concerns both the politics of identity and the politics of Nature.

5
Quarrying pebbles

The 2009 UK Saint Index of corporate reputations finds quarrying to be the most hated form of development in the UK (Saint Consulting Group 2009). There has been a long history of quarrying on the heathlands, provoking environmental concerns on the part of professionals and public alike. Recently local villagers joined with the Campaign to Protect Rural England, Natural England and the Environment Agency to oppose a bid by Aggregate Industries (hereafter AI) to mine for sand and gravel in the area around Straitgate Farm, which is a greenfield site in the north of the Pebblebeds, a site intended to substitute for existing operations near to the end of their life. In particular campaigners expressed disquiet about the possible disturbance to the underground watercourse and the effect this will have on the nearby ancient woodland and habitat, together with possible adverse effects on the flood risk for Ottery St Mary (Wright 2013).

In fact, Brown *et al.* have researched the economic contribution and environmental cost of aggregate extraction in England and conclude that indigenous supply is likely to be maintained in the coming years (Brown *et al.* 2011). AI's Corporate Environmental and Social Responsibility (CESR) practice could be described as eventually successful in relation to quarrying on the Pebblebed heathland in the long term. At an international level Hilson feels there has been considerable improvement in CESR practice, particularly in the areas of environmental management and community development (Hilson 2006: 225). However, the extractive industries' practices are often criticized, particularly by community-based NGOs and activist organizations (see Whitmore 2006; Curtis

2007; Public Eye 2011; Mines and Communities (n.d.); Foster 2010; Kirsch 2010a: 92, 2010b: 299).

During the eighteenth and nineteenth centuries there were numerous small-scale and shallow quarries across the heathlands providing building materials for local villages and farms. It was not until the early twentieth century that much larger operations began at Rockbeare Hill, Venn Ottery and Black Hill, where the exploitable pebble deposits reach their maximum depth of up to 30 m. Black Hill quarry was first operated in the early 1930s by the hand digging of pebbles and their breaking to create road macadam. The operations were rapidly mechanized with the installation of a crushing plant, and the quarry had its heyday from the 1950s until the 1990s when it developed into a huge operation. A massive extension of 57 ha was granted to the quarrying concession in the early 1970s following an application by CDE. This took place only a month later than the application to build championship golf courses on the heath (Chapter 2) and followed a failed application in 1968 to exploit an even larger area of heathland. The development, like the golf course, was contrary to Devon County Council's (DCC) own development plan but nevertheless was given permission. Quarrying operations eventually ceased at Black Hill in 2011 in tandem with the re-opening and extension of the disused quarry at Venn Ottery, last worked in the 1970s.

The quarries produce sand and aggregates for making ready-mixed concrete and building materials and high quality chippings for the surface dressing of roads. Its products are supplied throughout Devon and as far away as Sussex. Quarrying operations involve the machine-digging of the pebble deposits down to 5 m above the water table, up to about 25 m; the separation of sand and silt waste from the pebbles using water; and the crushing of the latter to various sizes and grades which are then stockpiled in huge dumps for future use.

The quarry produces five grades of crushed pebbles – 3 mm, 6 mm, 10 mm, 14 mm and 20 mm – as well as 10 mm, 20 mm and 40 mm rounds, together with coarse and fine sand. Of the material dug out of the ground on average 20 per cent is waste, between 40 and 45 per cent sand and the rest stone. Besides the crushing plant at Black Hill there are huge silt-collecting ponds, pipelines and pumps feeding 60,000 gallons of clean water to the plant per hour and removing the silty waste. The quarry produced 320,000 tonnes per annum in 2008. The pebble and sand deposits from the Venn Ottery quarry 9.5 km to the north and smaller amounts from Rockbeare Hill quarry are now transported to Black Hill for processing and crushing, involving a continuous stream of tipper and articulated

Figure 5.1 Part of the Black Hill quarry, aerial view

Figure 5.2 Sand tip and water-filled pebble extraction hole, Black Hill quarry

lorries moving across the western side of the heathlands along the B3801. In an average week there are about nine hundred HGVs arriving and leaving Black Hill and the tonnage produced was estimated to rise to 370,000 tonnes per annum in 2012. Material from the crushing plant is thereafter transported away, either to Rockbeare Hill quarry for macadam and concrete production or for use elsewhere for building and road surfacing. The grit produced from the crushed quartzite pebbles is extremely hard wearing and ideal as a surface road dressing. In a limestone quarry crushing plant blades will work for up to a year; at Black Hill the steel blades need replacing every six weeks, such is the hardness of the material. The pebble grit surfacing gives the rural roads of much of Devon their characteristic pink colour. It is also used for motorways. Pebbles have also been supplied from Black Hill for building walls in the locality. During the construction of the M5 motorway in the 1970s Black Hill pebbles were used to decorate the motorway bridges outside Exeter and 150,000 tonnes were supplied to build the new sea defences on Sidmouth beach.

CDE receive a royalty on the materials produced and the lease of land at Black Hill. Quarrying has, over the years, made a substantial contribution to estate income, turning an otherwise useless heathland waste into a profitable economic resource. In its early days the Black Hill quarry provided many local jobs but mechanization, including a new computer-operated processing plant, which was erected in 2003, had by 2009 reduced the labour force to just seven persons.

These quarrying operations raise interesting questions with regard to an area designated as an SSSI and AONB and in relation to recreational use of the heathlands. The worked-out Black Hill quarry is in the process of being restored and landscaped, work led by the Pebblebed Heaths Conservation Trust (PHCT) and the RSPB, who work with a plan drawn up by a team of landscape architects. This involves filling in the massive silt ponds, re-shaping the contours of the land surface, retaining shallow lakes and replacing a thin layer of topsoil on slopes to encourage heathland regeneration. The quarry and conservation managers are very positive about this landscaping project because it is creating a new wet heathland habitat complete with ponds and lakes, a type favoured over the original dry heathland with a pine plantation (in the original planning application from the early 1970s, now to be replaced) because of an increase in biodiversity. The claim is that the landscape has actually been improved by the quarrying operations and will be better than it was formerly. In some areas the quarrying has removed the concrete foundations of old Second World War military buildings, an additional benefit to the creation of new areas of wet heathland habitat. Areas covered with conifer plantations have also gone, to be replaced with a heathland habitat. The irony of all this is that planning permission would never be given for Black Hill quarry, sited within an SSSI, today, and the quarried area cannot be extended further. This sits somewhat uneasily with claims that the landscape has actually been improved by being quarried and bulldozed away, the original contours of the land gone for good to be replaced by artificial hill slopes and ponds. Members of Royal Marines training teams, horse riders, walkers and cyclists all commented on the manner in which the quarry had eaten away at an old, familiar landscape and transformed it, preventing access to a substantial area of the heathland which they had once enjoyed. Others take a different view:

> Wherever you start a quarry now nobody likes it, nobody wants it on their doorstep. We all want to live in a brick house, we all want to drive on tarmac roads but no one wants a quarry beside them. You can understand that but we've actually enhanced the area with the quarry. It's always good to see in hindsight. We've actually made that area better but nobody is going to see that thirty years down the line are they?
>
> (Assistant quarry manager)

This same line of argument applies to the renewed workings at the Venn Ottery quarry. Those undertaking the quarrying operations and also the RSPB and the Devon Wildlife Trust (DWT), who jointly manage three nature reserve areas directly adjacent to the quarry site, claim that it will enhance the environment in the long term at least. We might note that if such an argument was taken to its logical extreme all the heathland might be quarried away 'for the better'.

In August 2010 news that the Venn Ottery quarry, which had received planning permission as long ago as 1965, was about to re-open sparked local protest. Those opposed to the quarry argued that it would:

- have a devastating impact on the quality and diversity of the environment, affecting the breeding habitat of rare and endangered bird species, insects including butterflies, adders, toads and other amphibians;

- destroy a local landmark and visually impact upon the wider landscape:

  We have learnt that the work will also involve the destruction of the extraordinary stand of monumental Scots pine that sit atop Venn Ottery Hill and which have been a well-known landscape feature for many years. This area of woodland can be seen for miles around … and the work will be visible right across the beautiful Otter valley.

  (Protester)

- generate a massive amount of heavy road traffic down unsuitable country lanes, affecting recreational use by walkers, cyclists and horse riders.

The protestors claimed that the work amounted to the 'wilful destruction of one of East Devon's last really wild places' (Protest newsletter 2010). They pointed out that the old quarry site had once been considered for landfill by DCC. This had been ruled out by planners on the grounds that it would generate an unsuitable volume of traffic, lay within an AONB, and that the 'naturalized' quarry site was popular with dog walkers and local people. Now a far worse development had been agreed. They also claimed that the RSPB and the DWT had known for some time that the quarry was to reopen but had not made this public. Moreover, both organizations had benefited from considerable grants from the Aggregate Levy Sustainability Fund. In other words they had been bought off by the quarry operators and so were not objecting. The estates manager for Aggregate Industries claimed that 'we appreciate the short-term impact but we are looking to establish a 40-acre heathland site which we believe will be of great value to nature conservation. We shall be putting back a lot more than we are taking away' (BBC News, 31 August 2010).

On 24 September 2010 we participated in a public meeting to discuss the plans that took place in an especially erected marquee at the Venn Ottery quarry site. The meeting attracted a fair amount of local attention with a constant flow of visitors to the tent. An exhibition showing what work would be taking place, together with a variety of brochures and leaflets that demonstrated the environmental concerns of Aggregate Industries (AI), were freely available.

We first spoke with Richard, South West Operations Manager for Aggregate Industries, and asked him to explain their position in more detail. Richard reiterated the close relationship AI has had with the DWT and the RSPB both at Blackhill and Venn Ottery. At the latter site AI has assisted with the attempt to reintroduce the endangered southern damsel fly; has carried out gorse clearance, installed runnels through the landscape and undertaken some scrub clearance. He went on to explain how just over half of the site was going to be quarried and that the restoration process would be continuous. Once the quarrying was complete the site would be put into what he called 'after-care'.

When the issue of local concern was raised Richard said he felt that there was quite a lot of misinformation being spread. He said he did not know why this was the case and just then we were joined by local protestor Kyle, who was both articulate and vociferous in her complaints:

> I can tell you why, if you'll listen to me. You've known you were going to come here for a *long* time now. I was talking to a RSPB man over a year ago when they started putting fences up and asked, because I've walked this area, the fields, all over here, for twenty years now, and I said, 'What's happening over there? Is it going to be more RSPB?' and he said, 'Ooh, well, ah, you'll just have to wait for the surprise'. They'd obviously been told not to tell us what's going on because you've paid the RSPB a little bit of money … It's only in the last three weeks that people have become aware of what you're doing here and if you'd been less secretive and come right out at the beginning and said this is what we are going to do then you might, perhaps have had a little bit more trust from the local people who now probably don't believe a word you're saying.
>
> (Protestor Kyle)

She then went on to describe this open meeting as a Public Relations exercise, which Richard rigorously denied.

Kyle also pointed out that although letters were sent to the nearby village of Newton Poppleford nothing was given to local people who live much closer to the site and who were going to be adversely affected by noise, traffic and views. Richard then called over John Penney, who was responsible for the communications between AI and local communities; he explained that the Newton Poppleford parish council had been informed as well as the District and County Councils. After looking at a map he had to agree that people most likely to be affected did not live in the Newton Poppleford Parish boundary and that he would need also to inform Ottery St Mary parish and invite representatives to join a liaison committee that has parish representatives from different parishes.

Having been informed that AI is looking into erecting a formal bridle path through the landscape, which came as welcome news, Kyle went on to raise her concerns about articulated lorries driving on the single-track road, which has passing places some of which are not even long enough to contain the length of one of these trucks. Richard responded:

> Let me explain what we're trying to do with the vehicles. The vehicles are equipped vehicles. There are passing places and we're going to let the actual job of moving the materials to one company.
> (Richard, South West Operations Manager, Aggregate Industries)

Kyle asked him if this meant AI was trying to avoid responsibility for the transportation of the materials and Richard explained that AI owned no lorries and that using one haulier had several benefits. There would be only one point of contact; the drivers would be in radio contact and would warn each other if a car or other vehicle were approaching. He said that AI wanted the same drivers on the job every day so they would get used to the route.

> That's why we want to do it that way, because we've looked at all the civil impacts from this development and where the concerns lie. We're a good operator. I promise we will operate this with the least impact in the most responsible manner. That's what we are about.
> (Richard, South West Operations Manager, Aggregate Industries)

Richard then asked Kyle whether she would like to be shown round the site by him. An agreement of sorts was reached: 'Well, if you start opening another quarry, if you want to take my advice, you'll start notifying the local people a darn sight sooner than you have.' (Kyle) 'If we have not got the communications right on that one, I apologize to you for that, and, two, you know, this is all about communications; forget all the PR, yeah?' (Richard) 'I have.' (Kyle)

We then spoke with another protestor who has just moved into this area. She explained that she and her husband were aware of Venn Ottery being used for quarrying but had thought the lease had lapsed:

> Yes, we look on to Venn Ottery Common and we're concerned about what we're going to see and hear, and the dust, and the traffic.

They've taken this road over and it's now not even safe to drive let alone walk or cycle. Why are they ripping this apart? I don't see how in this day and age they can get away with it.

(Local protestor)

Seventeen months after this open meeting, after work has proceeded, we spoke with Kyle to see how she feels about the situation now. She said that she did not find it at all intrusive; they can see the lorries going from the quarry towards the Lynch Head Lane junction but they can't hear them; the only noise they can hear is when AI have earth-moving vehicles on site and there is a constant beep, which she believes are the lorries reversing. She continues to walk on Woodbury Common with her dogs and often finds herself in front of or behind one of the lorries on the single track road, but there have been no problems and the lorries have not affected her travelling. The formal bridleway is now in place, neatly fenced, and on the walking side there is no barbed wire (which was another of her concerns). A neighbour accepted an invitation from the AI Communications Officer to visit the quarry site. She has seen how the heathland is being reinstated when they have finished quarrying a section and is impressed with this and that AI have employed an Environmental Officer to oversee the restoration work being undertaken. Kyle says:

> So the lorries don't bother us at all but nonetheless I would say that I'm still glad that we did protest because we got certain assurances that AI have been very good at fulfilling. So by and large, and we're talking about the Venn Ottery Quarry, I have no problems with it at all. If we get in contact with John we can always arrange to go and see what's going on and how they are reinstating the heathland.
>
> (Kyle)

Conclusions

Quarrying in the Pebblebed heathland is a meeting between industry, local government and environmentalists. Those who live near the locales have a different perspective, one that through their protest has been recognized, with their views being integrated into the reinstatement of heathland and, in the case of Venn Ottery, reflected in the provision of a formal bridleway. Brosius argues that environment is both represented and claimed, constructed and contested (Brosius 1999: 277). It appears that the re-establishment of quarrying at Venn Ottery meets with this suggestion. But it is also interesting to note that in the example of Venn Ottery, the landscape is not just being constructed but re-constructed. Perhaps this will be reflected in how people will remember this landscape in the future, the new bridle path acting as a reminder to forget what lay here in previous years.

Milton describes how concern about nature protection encompasses ideas, feelings and practices (Milton 2002: 6). Regarding quarrying, this may be about not only nature protection *per se* but the response of people to what is perceived as a destruction of a scenic area; Milton gives as an example the Harris superquarry at Lingerabay, which would have resulted in the chiseling away of nearly a third of the magnificent mountain of Roineabhal on the Isle of Harris in the Western Isles of Scotland (Milton 2002: 135–46). Described as a 'tortuous tale of almost epic proportions' in what is regarded as Scotland's biggest ever environmental campaign, one lasting nearly 30 years (Scott and Johnson 2006: 3, 99), the Lafarge Redland Aggregates company eventually announced that they would not launch a second appeal against the refusal of planning consent for the quarry (2 April 2004), stating 'Harris is no longer a subject, now or in the future' (Scott and Johnson 2006: 129). What is of importance here, in relation to the Venn Ottery Quarry, is a further comment from Lafarge: 'This outcome will hopefully reflect helpfully on the company's positioning in a business environment where competition to raise ethical standards can offer competitive advantage' (Scott and Johnson 2006: 99).

Like Lafarge, Aggregates Industries (AI) too recognizes that it is vital to be seen to be concerned about issues a local community may raise, and, crucially, the community in opposition to recommencing quarrying practice at Venn Ottery are aware of how they may be able to use this 'business pressure' to achieve at least some of their aims.

Being able to be part of this landscape, walking, riding, appreciating its habitat, can be important features in an individual's life. Being able to express feelings for a landscape, what it means to one's life, has historically been disregarded in situations where a business company is intent on getting access to a landscape in order to exploit its natural resources, to make money out of it. A community's feelings are regarded as emotions and are of little or no importance. Increasingly, however, capitalist enterprises find themselves having to give more consideration to the needs and psychological well-being of local communities when they express their feelings, whether or not they are couched in 'scientific' terms, not least because of the media's interest in such matters.

Part II
The landscape as leisurescape

In this part we consider the ways in which the general public use and experience the heathlands. Chapter 6 presents the results of a very general questionnaire survey of heathland visitors. This is intended to provide a general background and introduction to the subsequent chapters. These consider in much more detail six different groups of people who use the heathlands for leisure activities on a regular and repeated basis: cyclists, horse riders, walkers, artists, people who fish, and model aircraft enthusiasts. We consider their activities and interactions both with each other and those who work in and manage the heathlands. Other user groups include an archery club and geocachers but there is no space to consider them here.

6
Introduction: the public and the heathland

A car park survey of fifty visitors to the Pebblebed heathlands was undertaken on three occasions (once on a weekend and twice on days during the summer holiday period: Sunday 5 April 2009 (17 informants) and Thursday 15 (10 informants) and Friday 16 July 2010 (17 informants). The surveys were undertaken by Kate Cameron-Daum and Chris Tilley, assisted by Jim Cobley in April. The survey was conducted in the most popular and best-known car park on the heathlands, at Woodbury Castle on Woodbury Common. Interviews took between twenty and thirty minutes using a structured questionnaire. Some of the questions were formulated with input from Bungy Williams (Commons Warden), Toby Taylor (RSPB), Tom Sunderland (Natural England) and Cressida Whitton (Historic Environment Division, Devon County Council). The questionnaires were undertaken anonymously. Thirty respondents (60%) were female and twenty (40%) male. Of these eight (17%) were under thirty years of age, twenty-nine (62%) were aged between thirty-one and sixty and ten (21%) were over sixty. Thirteen (26%) were educated to degree level, eight (16%) had 'A' levels or diplomas and the rest had GCSEs or no formal educational qualifications (32%). Occupations were wide-ranging, from a quarry worker to agricultural students, tourists on holiday, shop assistants, a chandler, science teachers, business people, members or ex-members of the Royal Marines, housewives, nurses, a retired professor, and members of the police and rescue services.

Table 6.1 Frequency of visits to the Pebblebed heathlands

Frequency of visits	Number of persons
Two to four times/day	2
Daily	10
Two to four times/week	10
Weekly	8
Monthly	4
Occasionally	12
Twice a year/yearly	1
First time/visitor	3

Visitor frequency

Informants were asked how often they visited the heathlands. Only three of them, all interviewed in July, were making their first visits. One informant was visiting the heath daily during a camping trip in the vicinity. Another visited about once a year; four came on a monthly basis with twelve visiting occasionally. Eight came weekly, ten came two to four times a week, another ten daily and two between two and four times per day (Table 6.1). All travelled by car. Five people, all interviewed in July 2010, were visitors to the area, coming from Wellington, Leeds, Norfolk, Plymouth and Northamptonshire. The rest (90%) described themselves as being local. Five were from Exeter and the rest from villages and towns in the vicinity of the Pebblebed heathlands – Exmouth, Budleigh Salterton, Topsham, Woodbury, Whimple, Lympstone, Newton Poppleford, Bicton, Farringdon and Clyst St Mary. The survey indicates that very few visitors from outside the local area visit the Pebblebed heathland. Sixty per cent visit on a very regular basis indeed.

Reasons for visiting the heathlands

The vast majority of these people came to walk on the heathlands (forty, or 80%). In addition seven either walked, cycled or came to run occasionally (14%). One had specifically come bird watching and two others for other reasons (yoga exercises and to 'chill out'). Twenty-five persons (50%) had specifically come to exercise their dog and for no other reason.

Dog walkers are thus by far the largest user group of the heathlands, some visiting on a daily basis. One respondent from Whimple, 10 km away, said he came four times every day with his dog.

Length of visit

Twelve of the visitors – all of them dog walkers – spent between ten and forty minutes on the Commons (24%); twenty three (46%), again mainly dog walkers, came for up to an hour, six between one and a half and two hours and nine for between three and six hours. Thus most visits were of short duration, taking place within the immediate vicinity of Woodbury Castle. Dog walkers therefore tend to stay for a relatively short period of time, anything between fifteen minutes and one hour. Other walkers without dogs, together with cyclists and horse riders, tend to stay longer and venture much further out into the heathlands.

Visits to other areas of the heathlands

Informants were asked whether they visited other areas as well as Woodbury Castle (Table 6.2). Twenty or 40% said they did so whereas thirty or 60% only visited this one area and most did not know the names of other areas of the Commons. Of the minority who visited other areas, Aylesbeare Common and East Budleigh Common were the most frequently mentioned.

People were also asked which was their favourite Common or area of the heathlands. Two had no favourite area, one stated East Budleigh, two Aylesbeare Common and seventeen Woodbury Common.

Table 6.2 Other commons mentioned by informants

Other areas visited	Number of informants
Aylesbeare Common	13
Bicton Common	10
Colaton Raleigh Common	9
Dalditch Common	7
East Budleigh Common	12
Harpford Common	7

Likes and dislikes

When people were asked what they liked and disliked about the heathlands there were a wide variety of responses. However, by far the greatest number mentioned that they liked it as a wide open space (seventeen, or 34%) and sixteen (32%) appreciated the extensive uninterrupted views. Other reasons for liking the heathland were far more variable and only given by up to eight (16%) or fewer different respondents (Table 6.3). These included mention of the many and varied paths, the fact that one was away from roads and people, that it was a variable and mixed landscape with plenty of interest and that it felt quiet, safe and peaceful. Only two respondents specifically mentioned birds, despite the fact that this is the principle reason for its designation as an SSSI. Four mentioned natural history or wildlife in general and only one the historical interest of the heathlands. As regards dislikes the majority of complaints were about dog mess not being cleaned up (22%); informants citing this included some dog owners. Seven (14%) complained about the lack of bins for dog mess and litter. Five complained about cyclists scaring dogs or horses (10%). Other comments were limited to just a few individuals. Twenty-one or 42% either did not respond or stated that there was nothing they disliked.

Forty (80%) of the respondents knew that the Pebblebed heathlands were owned and managed by Clinton Devon Estates. Others had no idea or thought the heathlands might be owned by the National Trust or some other conservation organization.

The archaeological and geological landscape

Thirty-seven respondents (74%) were aware that there were archaeological remains on the heathlands. All but two exclusively referred to Woodbury Castle, right next to the car park. However, only twelve (24%) knew that it was an Iron Age hill fort despite the fact that there are a number of information boards on its perimeter, one being in the car park itself. Twelve people were specifically visiting the castle (24%) while thirty-eight (74%) were not. Those who knew it was there found out about it through local or personal contacts (12%), the information boards (four, or 8%), through having trained in the Royal Marines or through knowing a Marine (two or 4%) or reading (another two or 4%). Three (6%) had found out about it by chance when visiting the heathlands.

Table 6.3 Likes and dislikes of visitors to the Pebblebed heathlands

Likes	Number of respondents
Open space	17 (34%)
Views	16 (32%)
Varied paths	8 (16%)
Mixed/varied landscape	6 (12%)
Away from roads and people	6 (12%)
Peaceful, quiet and safe	6 (12%)
Trees/woodland	5 (10%)
Good for walking/cycling	4 (8%)
Good for walking dogs	4 (8%)
Wildness/uncultivated	3 (6%)
Scenery	3 (6%)
Free car parking/good car parks	2 (4%)
Freedom	2 (4%)
Memories (childhood)/personal	2 (4%)
Roughness of terrain (for biking)	2 (4%)
Extensiveness	2 (4%)
Convenience of location	1 (2%)
History	1
Big skies	1
No sheep	1
People you meet	1
Heather/gorse	1
Pebbles	1
Reservoir	1
Dislikes	*Number of respondents*
Nothing/no comment	21 (42%)
Dog mess not cleaned up	11 (22%)
Lack of dog mess/rubbish bins	7 (14%)
Cyclists	5 (10%)
Potholed car parks	3 (6%)
Reduction in number of car parks	2 (4%)

(*Continued*)

Table 6.3 (cont.)

Dislikes	Number of respondents
Noisy model aircraft flyers	2 (4%)
Litter	2 (4%)
Presence of quarry	2 (4%)
Royal Marines firing	2 (4%)
Horse riders	1 (2%)
Youngsters	1
Adders and other snakes	1
Lack of signs	1

Thirty-four or 68% of respondents regarded the heathlands as being a natural landscape; only three (6%) knew that it was managed. Thirteen (26%) did not know. Thirty-one people (62%) thought that it was wild, fifteen (30%) did not know whether it was wild or not and only four (8%) thought that it was not wild. When asked if they knew it was a unique landscape of pebbles, thirty-three or 66% answered yes and seventeen (34%) no. People were asked to briefly describe what they thought about the pebbles in the landscape. Twenty-four (48%) had no opinion. There were a wide variety of responses from others. Four people said they were fascinating and that they had found out about them from information boards, a couple of others from the Pebblebeds Project website. Two found it quite amazing to be walking on an ancient river bed, three liked looking at and walking on the pebbles. One found the vegetation on top of them more interesting. A couple mentioned they lived with pebbles as they also had them in their gardens and on the beach. Two people mentioned that the pebbles made walking more interesting, and one that they made for more challenging bike riding. One cyclist disliked the fact that they made the track surfaces bumpy. For one person it was like being on the beach. Another mentioned the different colours and a third said she liked to massage her feet on them and noted that their beautiful texture made one feel part of the earth while walking.

Nature, conservation and threatened species

When asked whether they were aware that any endangered species lived on the heathlands, twenty-five (50%) of the respondents answered yes and

Table 6.4 Respondents' knowledge about the presence of endangered species on the heathlands

Species	Yes	No
Ground-nesting nightjars	10 (20%)	40 (80%)
Dartford warbler	16 (32%)	34 (68%)
Silver-studded blue butterfly	8 (16%)	42 (84%)
Blue damselflies	8 (16%)	42 (84%)

twenty-five (50%) no. People were then asked if they knew about the presence of particular species.

As Table 6.4 shows, a very low knowledge of endangered species is characteristic of people visiting the heathlands, the best-known bird species being the Dartford warbler. When asked whether they knew it was the nesting season for nightjars only seven people (14%) answered yes. Cyclists were asked whether they were concerned that cycling might cause damage, particularly if they went off the track and up and down the ramparts of Woodbury Castle, where much erosion has occurred, necessitating restoration works in 2009. Seven (14%) were aware of this while three (6%) were not. One answered that they thought it was important to stay on the tracks where there were plenty of humps and there was no need to go anywhere else. Dog walkers were asked whether they knew that they were advised to keep their dogs on a lead during the bird-nesting season. Thirteen (26%) answered yes and twenty-five (50%) answered no. But all but a few dog walkers were observed to let their dogs off the lead anyway. When asked about dog mess and whether they knew they should bag it up and take it home with them to stop nutrient levels increasing on the poor and acidic heathland soils only three answered yes (6%), six (12%) answered no and the rest were not sure.

Describing the heathland

People were asked to give up to ten words to describe this landscape. A few found this difficult and described it using a sentence instead. Of those providing descriptive words by far the most common was 'open', used by twenty-one (42%) respondents. After this the most frequently used words were 'beautiful' (thirteen, or 26%), with another three respondents calling it 'pretty', one 'astonishing' and one 'stunning'; 'varied' (thirteen or 26%), referring to a mixture of different types of vegetation, heath and woodland; 'natural' (eleven or 22%); and 'wild' (nine, or 18%). Six respondents (12%) chose 'heathland' or 'typical heathland', six (12%) 'green'/'greenery', five (10%) used the words 'peace/peaceful', 'heather', or referred to 'lovely' or 'nice views', and four (8%) said the landscape was 'unspoilt' or 'scenic'. So for the majority of visitors the heathlands are thought of as being an open, beautiful and varied natural or wild space consisting of gorse and heather heathland with trees; one that is green, unspoilt and scenic with lovely views. The responses did not vary significantly in relation to sex, age or socioeconomic class.

All other words or descriptive phrases were used only by between one and three persons. These can be grouped into five broad and somewhat overlapping categories as follows:

1. Words relating to how people feel being out on the heathlands and the effects the heathlands have on them, their body and how they feel emotionally:

    'calming' (two responses), 'lose oneself in it' (one response), 'spiritual' (one), 'trouble-free' (one), 'comfortable' (one), 'safe' (two), 'therapeutic' (one), 'freedom' (two), 'an escape' (one), 'nice to get away' (one), 'quiet' (two), 'fresh air' (two), 'private' (one), 'animal instincts opened up' (one), 'quite lonely' (one).

    These eighteen individual responses illustrate well the manner in which people specifically visit the heathlands as a means to escape from the stresses of everyday life and to cleanse and cure the body in various ways. The heathland acts as Other in relation to the rest of their lives, a place to which they can escape from the pressures of social relationships and responsibilities and achieve a different kind of experience not possible elsewhere.

2. Words relating to the manner in which the heathlands stimulate more specifically a more cognitive interest:

    'enjoyable' (two), 'exciting' (one), 'secretive' (one), 'interesting' (4) 'lots to investigate' (one), 'interesting variety' (one), 'adventurous' (one), 'rich' (one), 'ancestral' (one), 'mystical' (one), 'magical' (one), 'can feel the past here imagining how people once survived' (one), 'treading in forefather's footsteps thinking of Iron Age people making practical use of it' (one).

    These fifteen responses indicate another and quite different reason for visiting the heathlands. For some they are places where they can discover and experience something new in the present, and this primarily relates to the varied character of the natural environment. It enlivens the mind. For a minority this also specifically relates to a distant ancestral and prehistoric past associated with the heathlands and not with elsewhere that has a magical and mystical significance.

3. Descriptive words relating to the character of the terrain:

'textured' (one), 'cold' (one), 'feels like a desert, can feel dry' (one), 'barren' (one), 'subtle' (one), 'stony' (two), 'pebbly' (one), 'varied' (one), 'vast' (one), 'big' (one), 'large' (one), 'spacious' (one), 'expansive' (one), 'lots of room' (one), 'remote' (one), 'compact' (one), 'hilly' (two), 'not too hilly' (one), 'hilly in places' (one), 'up and down' (one), 'exposed' (one), 'starkness of the area' (one), 'rough' (two), 'rugged' (three), 'unique combination of sea and common' (one), 'windy' (two), 'water bits' (one), 'like Africa' (two).

These thirty-five answers are interesting responses to the experience of the landscape. They refer to its ever-changing character, the unevenness of the terrain; that it seems barren and rugged to some and very large indeed despite the fact that objectively the area covered by the heathland is no more than a few kilometres wide, and looks small on a map. Being there provides a quite different experience. Two respondents specifically likened it to the African savannah, a third as feeling dry and being like a desert. These are all, in their different ways, embodied experiences of this landscape.

4. Descriptive words specifically relating to vegetation and wildlife:

'bird song' (one), 'living' (one), 'yellow' (two), 'bracken' (one), 'wildlife' (one), 'wild flowers' (one), 'lots of plants' (one), 'moorland' (two), 'full of life' (one), 'woodland' (three), 'trees' (one), 'ancient trees' (one), 'oak scrub type common' (one).

These seventeen responses show the manner in which people respond primarily to the character of the vegetation while being out on the heathland. Only one specifically mentions birds as being an important part of the place.

5. Evaluative words relating either to the heathlands or activities possible on them:

'nice mixture of things' (one), 'unique' (one), 'valuable' (one), 'endangered' (one), 'well kept' (one), 'good paths' (one), 'good car parks' (two), 'good walking' (two), 'good for cycling' (one), 'easy walking' (one), 'extensive walking' (one), 'wonderful local area' (one), 'privileged to enjoy it' (one), 'interesting historical value' (one), 'well managed' (one).

These seventeen responses all positively evaluate the character of the heathlands for the kinds of opportunities or affordances they provide for people, the manner in which they are managed, and why they are important. There were no negative comments or descriptions and the responses emphasize the importance of this place primarily to people of the locality as a local resource.

Conclusions

Overall the responses emphasize the significance of personal embodied experience of the place. The heathlands are not felt to be important as a conservation area of national importance, or as a place where rare bird or insect species might be found, for the presence of particular kinds of archaeological monuments, or for their unique geology. These are all rather abstract ways of thinking about them. Informants' responses instead related to the personal possibilities they afforded for nurturing body and mind in a particular landscape set apart and away from everyday life experiences and interests. Key aspects of this were notions that this was a natural, wild and varied landscape and that it was open rather than enclosed and bounded. The fact that this was a rare example of a lowland heathland was not important. The supposed naturalness of the landscape was the important thing, and when people spoke of it being varied what most of them meant was the presence of woods and trees and vegetation that in an ideal heathland simply should not be there. The Iron Age hill fort of Woodbury Castle, for example, has old beech trees covering its ramparts and is adjacent to a contrasting conifer plantation. The survey was undertaken before part of the heathland was fenced off, and the responses reveal the manner in which its extensive and continuous management (as discussed in Chapter 2) is largely invisible to most people.

7
Modes of movement through the landscape: cycling and horse riding

In this chapter we consider two modes of movement across the landscape: cycling and horse riding. Both of these are mobile relationships in which encounter and perception unfold as part of a journey. As the journeys change so do these experiential and perceptual encounters. Unlike walking both cycling and horse riding are mediated, in one case by the technology of the bike and the manner in which it is ridden, in the other by the personal relationship between horse and rider. Both involve different forms of embodied action. These mobile relationships can be contrasted with those discussed in Chapters 10 and 11. For the fishers and model aircraft flyers the relationship with landscape is considerably more static. They return to the same place time and time again and develop accordingly a different kind of relationship with that place than do the horse riders and cyclists. They experience places sequentially, as do walkers, and their relationship with these places is thus of a different character.

Cycling: an embodied identity of challenge and pleasure

During the last decade there has been a resurgence of interest in cycling in the UK, although it is still the case that only a tiny proportion of the population actually cycle on a regular basis, as low as 1% of travellers (Cox 2015: 20). Cox and others have pointed out that there is an inverse relationship between the prevalence of cycling and its significance for the formation of social identities and meanings in relation to its practice

(Horton and Parkin 2013; Vivanco 2013; Lanting 2014). However, the heathlands constitute an important regional locale for cycling in East Devon, with the nearby city of Exeter being one of the six nationally designated Cycling Demonstration Towns in a project that lasted from 2005 to 2011. The heathland and the green lanes connecting the villages around them attract both local cyclists and others from quite an extensive area including individuals, families and cycling groups, who visit from Exeter, Sidmouth, the Haldon area to the west and the Axe Valley to the east. Some cyclists are not interested in off-road mountain biking on the heaths themselves but prefer to cycle the surrounding sweeping green lanes. Others are intrigued by what is frequently described as the 'challenge' of cycling on the Pebblebeds themselves. They often use the lanes to reach the heathland and it is these mountain bikers and their experiences that are the primary focus of the discussion below.

As Vivanco has noted there is a paucity of cycling studies in the anthropological literature, despite cycles' globalized production and use (Vivanco 2013: 9). Virtually all recent studies have concentrated on cycling in urban contexts and thus have not considered how cycling relates to an understanding of and interest in the rural landscape, one of the principal concerns of our discussion here. The literature that does exist is heavily dominated by practical policy implications; issues of urban planning and transport, safety, sustainability and social change, or is primarily concerned with cycling as a specific form of technology and its history. The bicycle has more broadly been regarded as a lens through which technological change, mobility and globalization may be thought through (Bijker *et al.* 2012; Rosen 2002; Urry 2007).

Anthropological interest in sport includes an initial structural-functional approach to games and play from the mid- to late twentieth century, and what transpired to be a more lasting cultural, symbolic and integrative approach; Geertz's (1972) writing on Balinese cockfighting as 'deep play', which he describes as a 'Balinese reading of Balinese experience; a story they tell themselves about themselves' (Geertz 1973: 448), is regarded as insightful and thought-provoking (Besnier and Brownell 2012: 445). However, Sands, who comes from a bio-cultural perspective, is not alone in arguing that until recently anthropologists have not been prominent in contributing to the study of sport despite the universal tendencies of play, games and sport helping to shape human evolution (Sands 2010: 5–6). Describing sports' influence, performance and meaning to different societies around the world as pervasive and unmistakable, he writes: ' … performance as it relates to the construction and maintenance of cultural and social markers of identity and reification of that identity, seems to be a thread that weaves through movement patterns in many cultures' (Sands 2010: 24). Dyck too notes the relatively scarce anthropological literature on sport but states that it is now growing. He also provides references to a number of interesting monographs on topics such as kabaddi, baseball and international football cultures (Dyck 2004: 4). Besnier and Brownell discuss how sport travels across boundaries and can illuminate issues within colonialism, globalization and sport mega-events, for example, and state: 'The anthropology of sport is now poised to make significant contributions to our understanding of our increasingly global society' (Besnier and Brownell 2012: 444).

Daring 'alternative' sports in which the natural environment, both terrain and weather, is integral to their practice have attracted some attention, including the anthropological. Such sports include rock climbing (Abramson and Fletcher 2007: 3), windsurfing (Dant 1998; Dant and Wheaton 2007), skateboarding (Borden 2001), bungee jumping (Cater and Cloke 2007) and tour cycling (Spinney 2006), as well as mountain biking. As part of his BA in Outdoor Leadership, Probert investigated whether mountain biking facilitates escapism and mental freedom and concludes that it does, with his respondents agreeing that they benefit from the therapeutic and challenging nature of this sport (Probert 2004: 40). Ethnobiologist Fowler provides us with an account of endurance mountain bikers as welding self to the landscape of Pisgah, in western North Carolina. These bikers travel down Farlow Gap, a highly technical steep trail over three miles in descent, where riders have to navigate obstacles such as fallen trees, loose boulders, creeks and waterfalls: 'Maneuvering a bicycle through Pisgah signifies skilful authenticity. Performing Pisgah defines selves and develops social networks – through praxis and narrative both on and off bicycles – while simultaneously exploring trails and deepening local environmental knowledge' (Fowler 2011: 11). Laviolette, who has participated in cliff-jumping, often known as tombstoning in the UK, writes that the attraction of such sports is attributed to the non-gratuitous individual confrontation of risk: ' … the dangers of personal injury being chiefly mitigated by honing physical skills and mental preparation … [which] helps [participants] develop personal skills in overcoming fear and managing risk self-reliantly' (Laviolette 2007: 1–2; see also Mauss 1979; Deleuze 1992). He then argues that such risks should be comprehended in relation to other human management of risky activities, including fiscal and sexual practices and environmental catastrophes.

In their interdisciplinary discussion of cycling and society, Horton *et al.* divide the cycling literature into four main areas: the historical;

the sociology of sport; the medical; and the design, engineering and planning perspectives (Horton *et al.* 2007: 8–9). Bourdieu writes that it would be foolish to assume that everyone participating in the same sport assigns the same meaning to their practice or even that they are undertaking the same practice (Bourdieu 1984: 209–211). This is certainly true of cycling, of which there are a number of genres, including those of cycling as a sustainable global means of transport; competitive cycling; leisure cycling; and the more 'alternative' sport of off-road mountain biking, the latter of which is the focus of our discussion on biking on the heathland.

Rosen states that the technology of mountain bikes developed as a result of biking enthusiasts who wished to ride where conventional bicycles were inadequate (Rosen 2002: 133), and it is not surprising to find that when it comes to mountain biking there is a range of literature concerning the prevention and management of physical risk. Besides the debate over the wearing of helmets, which has a research foundation devoted to it (the Bicycle Helmet Research Foundation), other recent papers include a discussion of other risk factors such as perineal trauma (Zandrino *et al.* 2004) and facial trauma (Kloss *et al.* 2006), a large-scale detailed assessment of mountain biking-associated spinal fractures and spinal cord injuries (Dodwell *et al.* 2010), and studies of the physiological demands of downhill mountain biking (Burr *et al.* 2012) and injuries sustained therein (Becker *et al.* 2013), which although dealing with a more extreme variation of mountain biking, also applies to those encountering rough downhill terrain on the Pebblebed heathlands.

For some heathland riders a major interest includes the model of the bike, how this can be adapted or replaced over time and the accessories for both bike and body. There has been much anthropological interest in the material culture of 'things', particularly approaches that articulate their meanings and social relations (for example, Miller 1998; Tilley 1999; Buchli 2002; Küchler and Miller 2005; Tilley *et al.* 2006; Naji 2009; Douny 2011). As yet a lengthy study of mountain bikers, their bikes and attire has not been written but there is certainly scope to explore a number of approaches; Warnier's praxeological approach, for example, could include factors such as the sensual and emotional; the memories each bike may encapsulate for the rider; the creation of identity that has been shaped, perhaps, by the embodied material culture of the biker; the transforming of space into a place embodied in movement; and the meaning silently conveyed by choice of bike and clothing worn. Artefacts can possess a silent form of communicative agency: 'It follows that without an exploration of the metaphorical power of things and the effects that these things have on people's lives we cannot adequately know or understand ourselves or others, what makes up our identity and culture, past or future' (Tilley 1999: 25).

Our interest here is in the relationship between people and their bicycles, their mutual interaction, and how this relates to a mobile relationship to the landscape: what bicycles do for people and what people do to them. This is an embodied relationship in which the bicycle becomes part of the body of the cyclist. In relationship to the landscape this involves ways of knowing, sensing and interacting with the world that differ significantly from those experienced by other groups such as walkers or horse riders. This general theme has been brilliantly explored by Borden in relation to skateboarding in the architectural spaces of the city (Borden 2001), but little explored in relation to cycling in either urban or rural contexts, beyond general comments about bicycling involving complex interactions between physical, social and experiential dimensions of movement (e.g. Vivanco 2013: 8; Cox 2015: 20). We start by considering the bikes people ride and their attitudes to them and their apparel. We then consider who rides with whom, before considering heathland relationships.

Mountain bikes and riding apparel

Several of the biking enthusiasts who use the Pebblebed heathland have more than one bike, including mountain bikes (where stability is the priority) and general road bikes or racing bikes where speed is favoured. Other bikers retain the same frame and change different parts of the bike, either through choice, such as to fit a different kind of handlebar, or for reasons of maintenance:

> Your chain, and the rear cassette (where the different gears are fastened together), because it's multiple gear, needs regular changing, perhaps once or twice a year depending on its use, and the clogs on the front chain wheels have to be replaced periodically. The wheels need replacing because you wear the wheels out when you break stones and that sort of thing. You grease the head set where the handlebars connect with the front forks; you grease the shock absorbers; you constantly buy new tires and inner tubes and new pumps and saddle bags and kit.
>
> (Jack)

Another rider who has two bikes, one of which is eighteen years old, the other ten years, says:

> Other people buy a new bike every year and in part that depends upon your view of it. I mean in my view it's more about the sort of physical capability of the person rather than the bike; the bike's almost irrelevant in many ways whereas other people want to have the latest and lightest and so forth.
>
> (Kirby)

He feels that fashions come round and describes how the top tube of the bike always used to be horizontal, until about fifteen years ago ones that were diagonal were produced – 'which means you had more clearance when you got off the saddle'. There is now a return to the horizontal tube, which he describes as creating a strong triangular shape. He goes on to say that 'serious women bikers' actually ride men's bikes because they are lighter for a given strength, rather than the bikes for women, which tend to have thicker tubes to support their more open frame and shape'.

The apparel worn by the riders may vary; for example, mountain bike helmets have a peak on the front – 'to keep mud out of your face really' – whereas the on-road bike helmets are without this feature in order for their riders to be more streamlined and dynamic. Obviously, the weather determines what the rider may wear, usually between one and three layers of clothing. Currently amongst racing bikers there is a fashion for black clothing, which other riders feel is foolish: 'I think it's terrible because it makes visibility, certainly when it gets dark, very, very hard' (Kirby). This biker rides both on and off road and as he and his group come from a road background where tights are worn, he and his co-riders wear black tights and brightly coloured tops: 'It looks a bit bizarre actually because we're all sort of getting on a bit and seeing these elderly gentlemen and ladies in tights, er, I'm sure it gives some people a bit of a shock'. He says that many people who do a lot of mountain biking wear fashionable things, the current fashion being baggy shorts with brand names such as Howies and baggy T-shirts. His tights are made by Altura and his jacket from a firm called Gore: 'They use Gore-tex material so you don't sweat too much'. Some of the mountain bikers also put on extra pads on the knees and shoulders to give greater protection.

There are four main areas in which differences between mountain and road bikes may be found; these are their shape, weight, tyres and suspension. Mountain bikes have wider handlebars that provide the rider with greater control than do the curved handles of many road bikes, which are lower and aerodynamic, hence the hunched position of the road biker, positioned closer to the top tube and the pedals. The mountain bike tends to be heavier and the rider more upright, with the bike possessing wider, knobbly tires (an increase in friction and surface area provides more stability) in comparison to the road bike's smooth, thin tyres. Whereas racing bikes may have no suspension, the mountain bike is constructed with features that absorb vibrations such as front shock absorbers and possibly rear suspension, which are useful when riding on the Pebblebeds. Front suspension means that when the handle bars are leant down on, the handle bars move down four inches: 'So if you go over a bump that's only two inches high, you don't get a jolt on your hands and the suspension absorbs the shock' (Kirby). Mountain bikes without rear suspension are known as hard tails, which seems an appropriate description – 'You get less of a hammering with full suspension' (Colm). The bikes with suspension front and rear are of higher quality and are more expensive. The range of suspension can vary also: 'Cross country bikes generally have about four inches of suspension and some of the big jumping bikes have six, seven, or eight inches of travel but they're also heavier' (Colm).

It seems hard tails and full suspension mountain bikes (MTBs) require different road techniques too because of the more shaky ride on the former but both types require certain alignment skills. 'You learn to stand out of the saddle when it's bad … the ground is unforgiving so you have to be prepared for the bumps and twists and the turns and to avoid obstacles' (Jack). Going downhill changes the centre of gravity and there is the danger of allowing too much weight on the front wheel, which can lead to going over the handlebars when braking. The technique here is to stand with the bottom behind the saddle and this adds weight to the back wheel. The legs are also used as springs to absorb the shocks of individual bumps: 'There are some bits where the bike actually jumps … when you're doing a drop the secret is to actually lift the front wheel so you fly through the air and your front and back wheels touch the ground at the same time' (Kirby). If the drop is steep and the landing is just on the front wheel, again, a somersault over the handlebars is a likely consequence. Alignment when riding uphill also involves standing when the hill is steep: 'As you go uphill there's a tendency for the front wheel to lift off – as you pedal harder it lifts the front of the bike up' (Kirby). In instances like this the front of the bike is therefore lighter, which means steering is affected: 'The bike won't follow the wheel and you can easily fall off so you sometimes have to move your body forward over the handlebars, lift your bum off the saddle, and stand up, so to get that weight on the front wheel so you can actually steer properly' (Kirby). One cyclist describes the importance of looking ahead and anticipating: 'Particularly when going through trees and that sort of thing, you know. What's very important is your hands on the handlebar; it would really hurt against the trees, so you've got to be careful' (Jack). Some riders find areas on the heathland that are so steep that the rider has to get off and push – 'there's no choice' (Kirby). An interviewee who takes children riding on the Pebblebeds says she reaches a particular gulley and thinks:

> 'Oh gosh, we're here again', and we've gone quite a trek to get to that point but, like, it's easier to go forward than it is back and luckily I always have helpers. A couple of the strong ones lift all the bikes up the gulley and I've got all these little children scurrying up there, they love it, they love the adventure.
>
> (Sam)

On occasions, a certain combination of weather and terrain also means pushing rather than riding:

> You've got more control when it's been raining but of course it never rains but it pours and it means the peaty areas become very sticky and slow. You sink, oh yes, the tire goes under the peat. The peat is quite a strong texture and if you've got a good cycle tire, most of the time you can pedal through, but sometimes it's so watery you just find yourself having to get off and push.
>
> (Jack)

Others find the need to lift their bikes over or up terrain. And so the experience of cycling on the pebbles can vary according to the weather.

> When you've got what I call temperate weather, not too wet and not too dry, it is best. If it's very wet, everything becomes very sticky, although the pebbles stay where they are, but it's more difficult for everything could be sticky, even the grass drags. When it's very dry, the pebbles become loose and so you tend to get a skid. There are a few steep descents and there are some I just don't do when it's dry because the stones will just fly around and you just haven't got any grip.
>
> (Jack)

When describing the conditions of the terrain, another rider remarks on how it only takes about two days for the mud to go if it has been raining: 'The good part about Woodbury Common is the fact that because of the Pebblebed heathland it does dry up quickly compared to other places, it runs off the stony pebbles' (Colm).

Riding groups

Some of the riders have been cycling on Woodbury Common for decades, one rider training there from when mountain bikes (MTB) were first

Figure 7.1 Riding group out on the heathlands. Courtesy of Chris Warburton, Knobblies Bikes

Figure 7.2 Riding group. Courtesy of Chris Warburton, Knobblies Bikes

introduced from the United States in the early 1980s. Other early users include the Sidmouth Valley Cycling Club (SVCC), which was formed in 1991 and has about forty members. The SVCC used parts of the heathland to promote race competition events consisting of five circuits, with a focus on the Woodbury and Colaton Raleigh Commons. A trail was marked out beforehand with flags and bunting and started from Four Firs car park; the organizer states that as the area of the quarry kept changing, they had to modify the route, particularly near the start of the race, to take in these changes and keep the riders away from any dangerous operations. It would take the more experienced riders just under ten minutes to complete a circuit, so they would be racing for about forty-five minutes before the finish. Trends have changed, however, and events are no longer organized on the Common by the SVCC because many of the current members now prefer road cycling, but individual members still participate in small social MTB cross-country rides. The SVCC racing competition remains well remembered by local cycling enthusiasts with some cyclists speaking of it as the first MTB race they had ever seen.

One cyclist we met is a former Royal Marine who took over recruit training at Lympstone in 1984: 'I had in any one year a thousand recruits going through Lympstone; sixteen troops of fifty and they would use the Common for a lot of their training. In fact if the Common were closed to the Marines, the Marine camp would have to move. Woodbury Common is the best training'. He retired from the Marines in 1989 and was introduced to mountain biking in 1990 and, as he puts it, has never looked back. He still goes out on the Common about once a week on a Friday afternoon and although he also rides on Dartmoor, the Quantocks and the Cotswolds, he regards Woodbury Common as being as good as you can get: 'It's not a large area but my goodness me it's a much easier and interesting place than Dartmoor, which can be very rocky paths and so on. Woodbury Common's got it just right'. As chairman of the Cycle Touring Club Exeter (CTCEx) and the organizer of their off-road rides, he takes them out periodically during the year on long day rides on Woodbury Common.

The thought of mountain biking across the countryside can feel quite daunting but the Exeter Mountain Bike Club (EMBC) runs several programmes that cater for all age levels. There are Confidence Builder rides for adults, a women's riding group and Go-Ride skills sessions for children and young people aged six to thirteen run from the Haldon site. In the summer months their coach brings the children to ride their mountain bikes on Woodbury Common and she also meets with the women cyclists there at least four times a year. For the adults, when confidence and skills have been built, there is a Wednesday night group that meets at the Four Firs car park at least once a month for a ride on the Common. The Wednesday group is a mixed group mainly consisting of men, some of whom are 'hardcore' members who ride their MTB every Wednesday evening, in all seasons, in all weathers.

Another group of mountain bikers who ride fairly regularly on the heathlands are the Axe Valley Pedallers who, based in Seaton, bike on the heathlands about twice a month, either in the evenings or at the weekend. They ride throughout the year: 'In the winter it's great because it's more to do with night-riding and then in the summer there are longer evenings to enjoy being out and about and it's always great' (Kimmo).

'Coffee Pot Rides' and the 'Bike Bus' are two other very popular organized cycle rides that often take in the lanes that traverse the Commons. The 'Coffee Pot Rides' recently celebrated their thirtieth anniversary. The organizer chooses a café in East Devon and usually between sixty and seventy people cycle up and meet there. Two years ago, when one of these riders noticed that although the meet-up itself was very sociable, the getting-there and return journeys were usually ridden on one's own, he set up the 'Bike Bus': 'I realized that there were quite a few people who didn't go on them because they either didn't know the way or couldn't mend punctures so I started the Bike Bus' (Kirby). A timetable is sent out every Tuesday evening showing the Thursday's route: 'People just join in on the way' (Kirby). With a mailing list of over 140 cyclists, there are usually between twenty and thirty cyclists taking part every Thursday. No one is ever left behind – if someone suffers a puncture, another rider will wait with the stricken cyclist until the puncture is repaired and guidance is also given on how to mend punctures. Despite an article in a cycling magazine that stated some women preferred women-only cycling groups, when asked, the Bike Bus female participants said they preferred to ride with men, and there is a 50:50 gender ratio in this club. The organizer comments:

> We tend to be very supportive in terms of getting people to ride, particularly on Thursday but on Tuesday (Coffee Pot Rides), quite to the contrary, we actually have a ride where we say if you get dropped you've got to find your own way back; if you have a puncture you've got to sort it out yourself. So we don't actually stop for anyone and if they have a puncture, it's up to them to get back; they know it's going to be a more aggressive ride whereas on the Thursday it's supportive and, you know, they make that choice.
>
> (Kirby)

In this way people of differing abilities and motivation are catered for. A community has been developed around the Bike Bus and one female member suggested the club create its own shirt. 'The design of the shirt includes the Devon hills so it's got rounded brown hills, it's got the River Otter in it and it's got the beach huts at Budleigh, so that kind of reflects the landscape, and that's become part of the Bike Bus' (Kirby).

Of course, some bikers also ride alone or with just one companion and find this to be a different experience from riding in a larger group: 'If you're in a group you obviously tend to talk to other people and sometimes you hardly know where you've been because you've been concentrating on what they've been saying whereas if you're on your own you're obviously experiencing things much more strongly' (Kirby). It is interesting here to compare this with those who participate in organized walks, such as the Ramblers, where a similar experience is shared. One cyclist who joined the CTC in Exeter describes how going out in a group is a fantastic experience: 'It's like going for a holiday every Sunday really' (Pamela). Others find that a leader making you go a particular way is not always what they want. 'When you're on your own you're looking around and in some ways it's easier to stop as well because if you see something interesting, like a broken tree trunk or

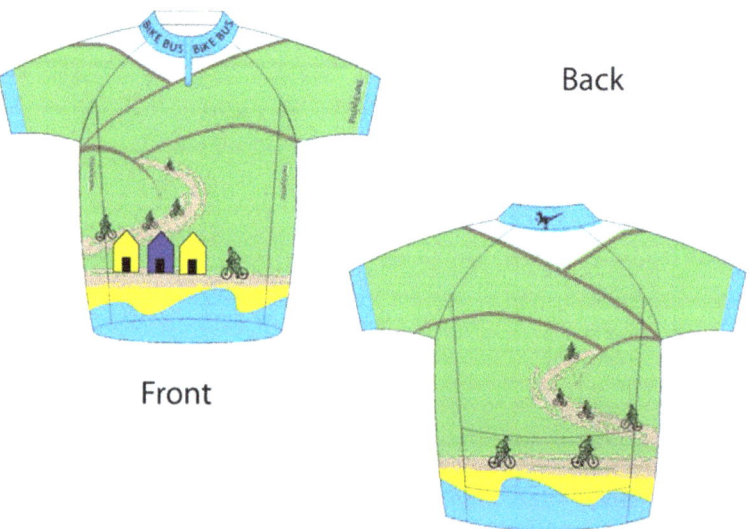

Figure 7.3 Bike Bus T-shirt, designed by Stephanie Houghton

something, you can stop and have a look whereas if you're in a group you tend to have to keep up and not get left behind' (Kirby). Another rider describes how when he is on his own he decides he will go and cycle for an hour or two, just wandering around seeing where the paths go, but when he takes friends onto the heathland it is not such a casual experience: 'I know the Common quite well so I've taken some friends from work around and I had more trouble with that. If you're taking people with you you're a bit more aware of not getting lost, of going somewhere that's more difficult to get back from' (Paul). However, one cycle leader remarks how if you go out as a group and you know what to look for, you can stop, and in the right places, learn about what is special about the heath: 'That's the beauty of living in a place, that you get to know it really well, you know the people who know the heaths really well, know what to look out for' (Kimmo).

Routes through the landscape

Those leading the group rides, such as the Bike Bus, tend to plan their route in advance, but others without such a timetable do this also – although some changes may later be made: 'Sometimes towards the end of the day I think, "I'll shorten it a bit" or "I'll lengthen it a bit"' (Jack). This particular group leader finds it surprising how some paths have disappeared whereas other new ones have appeared: 'I'm still finding new paths (and this after twenty-one years of riding the Commons). I go along a path I've used perhaps two years previously and it's overgrown but not far away there's a new route. It's a constant change all the time' (Jack). He feels that one reason new paths are created is that when a path is used a lot the surface tension is broken, resulting in a muddy area that remains waterlogged for a long period of time and prompts people to go round it. Another cause is when a route is not used very often; it becomes overgrown with gorse, and when other people start wanting to go in that direction they go round the gorse and a new path is created. He notes how volunteers cut away gorse from tracks that are used a lot and refers to the rotation in bracken-cutting too, comparing this management with that of the Brecon Beacons where he also sometimes cycles: 'The bracken's just been left and you just can't see the trails, I mean, you know, it's completely overgrown there' (Jack).

One rider with a great interest in maps has explored many of the green lanes and enjoys sharing these discoveries with other riders. He does say, however, that for mountain biking on the Commons, maps aren't necessarily a good guide: 'Some of the better tracks are, you know, hidden away, and not particularly well-marked' (Kirby). Another says, 'You learn where you can go', and describes how he tends to keep near trees in the winter when it is cold because there is more shelter from the wind: 'That's another reason to go round the edge rather than straight across the middle. I mean the Common is quite wet as well. There's lots of low bits where you can't go, so it's quite difficult to go across it' (Paul). He himself varies his route, sometimes approaching from what he calls the 'top' route, from Woodbury Castle, at other times going round the bottom of the Common from the Exmouth direction, past the quarries and coming into it: 'There's several different ways you can go in … You go right up against the edge of the quarry and it's quite nice starting off, it's all downhill and you go quite quick and it's quite a wide track so if you meet anybody walking you don't tend to hit them' (Paul). He goes on to remark that a new track has been put in here by the quarry and he does not particularly care for it: 'I don't really like that one; it's a bit open and boring and it also keeps getting washed away so there's some big holes. I had some friends hurt themselves down there' (Paul). And so expediency is also a factor. Another rider often cycles across Colaton Raleigh Common but avoids that part marked as a brook because it is so marshy: 'There are more tracks than are shown on the map but you're giving yourself unnecessary difficulty going there. It's much easier to go round and do more elsewhere than just waddling across here' (Jack). He speaks of the narrow tree-lined paths on the edge of the Colaton Raleigh Common and of when a linear copse seen when coming in from near Upbury Lane, probably 20 to 30 yards wide and even wider in parts: 'It's very attractive and there's a peaty part that runs all the way through there. I find that it's one of my favourite stretches because you're on a leafy path in the summer, that winds through the trees, but if you go off path you'll hit a tree', he says, chuckling.

Maps produced by cyclists not surprisingly emphasize a combination of surrounding roads, towns and villages, those crossing the heathlands, and favourite off-road tracks. One shows a wide area of the southern part of the heathlands as far east as the River Otter, along with twenty-one named places, a favourite place for bike jumps on Bicton Common, individual landmarks such as Hayes Barton house (birthplace of Sir Walter Raleigh) and local cafés.

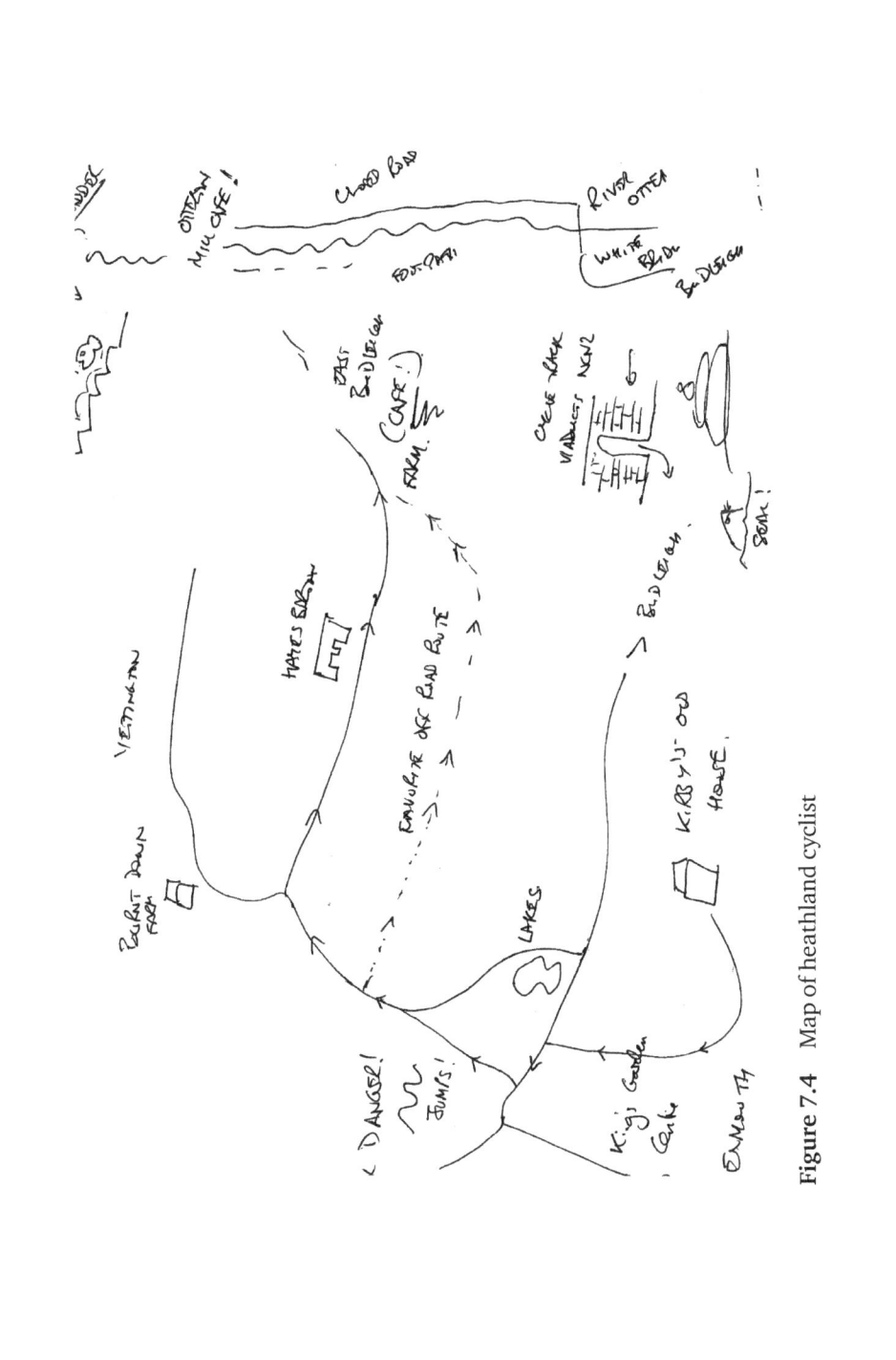

Figure 7.4 Map of heathland cyclist

Night riders

Riding at night on the heathland is also very popular and we discuss this next. The essential piece of equipment is a high-quality night light – 'the brighter, the better' – and these can be very expensive, one rider describing his light as costing more than his bicycle. He has a Venture Enduro model that lasts for three hours on full beam and clips on to the bike, and he states that the technology has changed a lot during the last five years: 'I had an old lead acid battery to start off with and I couldn't keep up with people and I realized when I got this one it was because I couldn't see. This makes a huge difference' (Paul). Some riders wear a light on their head, others attach one to the front of the bike and some use both. The choice of which kind of light to use can be a matter of what the rider can afford, the lights attaching to the head being more expensive – 'I can't afford to get the one to go on my head yet' (Paul), or it may just be a matter of preference. There are actual differences in the effect of what can and cannot be seen when using a head or bike light, and also the riding experience. The Venture Enduro has a 35-degree spread, which allows the biker to see far enough when going fast with sufficient time to brake when necessary, but another rider who also currently uses a bike light explains: 'What you lose with the light being on the bike is that you can't turn your head to see … you get peripheral vision if you do that' (Colm). However, as one's instincts are to ride in the direction of the light, when the light is attached to the head rather than the handlebars, and the head is turned to one side to look at something, the biker may also cycle off in that direction and come off track or have an accident. Yet other cyclists prefer the light attached to the head as it apparently helps the rider to focus more and is brighter. Although she regards head lights as preferable for these reasons, one rider uses lights attached both to her head and to the bike handlebars when she is coaching children in mountain bike riding, as she needs sufficient light in case of emergency to look after everyone and get them all back safely.

One hazard is when turning round and becoming dazzled by the light of the rider behind you. The Royal Marines have an issue with the dazzling effect of the cyclists' lights too; their use can cause confusion during manoeuvres, as one rider explains: 'We bump in to them and they've got their lights on and we've got our lights on … we dazzle them and they don't really like that … they do complain every now and again' (Cohen). The Marines are not the only ones who can be dazzled: 'We see lots of rabbits up there and they can be a bit of a pain actually because they just stand there and look at you because they get cornered by the lights and I've had a few near misses because they don't run' (Colm). Other nightlife on the Commons is also affected and the riders call out warning to each other: 'Frog, frog', for example. Owls are often seen and sometimes 'spooked' by the riders:

> And of course if you get them spooked they then tend to spook other animals up there like deer; it can frighten you to death because you're riding along and this deer will jump out from anywhere and run off; that's quite spooky. You know, they're there, sort of thing, and then, phwor, they're gone.
>
> (Colm)

Differences between day and night riding

Many cyclists do not ride at night and the reasons for this are various, including the cost of decent night lights, a fear of disorientation, and a preference for seeing where one is going. Levels of fitness and confidence may also be factors when night riding. One cyclist spoke of his caution when deciding to ride at night and of how he only went out by himself when first starting, as he was unsure as to whether he would be able to keep up with the experienced night riders who rode in groups. He now rides on a Tuesday night with the Exeter MTB Club and they often ride on Woodbury Common. Another said it was good when the RM were out: 'It's always quite nice and comforting actually sort of knowing they're around really' (Paul). Also, two riders describe their Thursday night ride as not being a club as such, chuckling: 'It's a kind of a mates thing, it's by word of mouth and if people think they're fast enough we'll invite them along' (Cohen).

The cyclists have spoken of their pleasure in riding the commons because they are so open and the views are beautiful, but what is it like at night when the views are no longer visible? Certain riders find that it is more of a 'heads-down' experience because there are no views to be looked at, and describe it as being a 'different ball game': 'You can ride a trail in the day and when you go back at night it's completely different', said Colm. He feels this is because the trails are ridden more slowly and the riders' perceptions are different because of the darkness. Nevertheless, the actual cycling still appears to follow a pattern of a burst of activity and then a stop – to wait for the slower riders to catch up, a pause to discuss which trail will be followed next, or a wait whilst someone fixes a puncture, and stopping just to look up at the stars. Does it still feel 'open'? To some it does not feel this way; the feeling of spaciousness is lost and the senses may be used differently because the rider is more alert: 'It smells different and your eyesight has to be much better because things come at you much faster because you just don't see them coming' (Paul). The sense of hearing also appears to be particularly important at night: 'You need to respond quickly to the sound of what's under your tyres, you know, from the point of view of whether something's loose or solid or wet' (Kimmo). One rider feels closer to the landscape at night because he feels as if he is moving in his own space – 'It's like a little sphere travelling with you' (Paul). He finds this quite comforting in what he describes as a spooky landscape: 'It's quite warm and friendly. It feels smaller and more like you feel when you're at home, nicer somehow because you feel more at one with it but you've got to treat it with respect' (Paul).

Comparisons with walking

Several cyclists make interesting comparisons between walking and biking:

> Well, I've done both and I enjoy each for its own sake. Cycling is more challenging, physically more tiring, it's exciting, and it helps to develop a degree of fitness. You can invent challenges like steep hills and sudden drops and so on as well as seeing the scenery and, you know, cycling with the CTC, you've got the route, reconnaissance, you've got to find your route, you've got the challenge, the physical challenge, you've got the company of the people you're with, that's three jolly good reasons, isn't it, and you've got fitness.
> (Jack)

He says he sees more when walking rather than cycling but, because bikes are very quiet, he often surprises animals such as foxes and badgers while cycling quickly and quietly: 'If you're talking when you're walking then you won't see much, you've got to walk quietly, but if you're cycling and you're quiet, you catch up all sorts of things' (Jack). Another cyclist remarks upon coming across a hare on one occasion and feels he would not have seen it if he had been walking. Others find cycling more rewarding than walking because a greater distance is covered and it is more of a challenge.

Some mountain bikers prefer this activity to walking or on-road riding:

> Yeah, the big advantage of the mountain bike (as opposed to an on-road bike) is you can just go for a gap; whatever the terrain. It's really nice, nicer than walking in that you're higher up so you can see over hedges, you can see where you're going to be and where you've been and you can go a lot further. I mean I would never dream of walking from here to here in an afternoon, and come back again but on a bike it's not difficult.
>
> (Paul)

He feels that he sees more of the nature immediately around him when he walks but prefers to do mountain biking as he sees the landscape from different perspectives: 'You don't have to be cycling fast all the time'. However, one cyclist who also does a lot of walking on the Commons says that she does not have a preference between walking and cycling as she gets a different experience every time she is up on the Commons. One of the cycle leaders also comments on speed differences when walking or biking: 'You're going to be moving at a completely different speed so it's fair to say that, walking, you will see more, up close, and are more likely to see things than you would on a bike' (Kimmo).

Relationships with the heathland

Cyclists discuss their feelings about the heathland and whether they find it to be a natural wild place. As will be seen, there are commonalities and differences among their viewpoints. One cyclist describes his feelings when on the Commons thus: 'It's a feeling of freedom and challenge I suppose. And I find it's nice to be there. It's nice to be free and forget the worries of the world and your age and things. It makes us young again and off you go. That becomes more important with time', he says, laughing. He has no particular favourite route: 'I just enjoy being there, whatever it is. I enjoy doing a long day rather than a short day because you've got that greater feeling of the whole day of freedom and just doing what you want to do. But that's early, I mean I like going off early in the morning and coming back sort of late teatime' (Jack). He says, when walking, he senses being part of nature but he also feels this when cycling too: 'I mean we're not going flat out all the time; I make a point of stopping and looking'. He does not feel the Commons to be a wild place, however, as 'wild' to him is walking in the wilderness in North America.

Another says he frequently goes out on his mountain bike on the Commons and now knows it like the back of his hand: 'So I picture where to go and I can work out how long it's going to take me to go everywhere but conditions (of the terrain) make a huge difference. It's just nice to be in the outside world and away from everything' (Paul). He describes Woodbury Common as being a fantastic place for mountain biking because it has everything: 'There's flat and open wide stretches where you can go for quite a long way and there's a single track, which is narrow and twisty, in among the trees and I like that, you can get up some speed and it's fun'.

When he comes across paths with obstacles such as fallen trees, he removes them. He also comes across paths with holly growing across and says these can be quite hazardous, particularly when biking in a group because one can easily get holly in one's face, and so he sometimes takes secateurs with him to remove such obstacles. He laments the cutting down of trees because although more of a view is revealed when they are cut down, he feels there is a loss of certain wildlife. This cyclist likes to be quiet and see what is going on around him but he rarely feels part of the landscape: 'I always feel humans intrude on my view of the world' (Paul). Although he views the contours of the landscape as being natural he knows the Commons are managed:

> It's obvious that it's managed because there are paths everywhere and you can see where there has been intended burning and when it is accidental burning. You can see where they have been mowing down the gorse and you can't go very far without seeing the quarry or bits that they have worked on. So, it's natural in that it's full of nature, wildlife and stuff, but it's very managed.
>
> (Paul)

And another rider states he does not feel the heathlands is as natural a place as it once was due to the expansion of the quarry: 'Where the quarry is now we used to ride trails down through there and they've dug it all away' (Colm). He finds that the trails grow over very quickly unless the growth is managed or there has been a fire. He has the sense of the heathland as feeling wild at times, although not as wild as Dartmoor, Exmoor or the Quantocks.

Mountain biking on the Commons is often referred to as having bursts of activity and this is felt to be good for health reasons. The physicality of riding is pleasurable, as is following the contours of the landscape and anticipating what is coming next:

> If you don't anticipate you certainly get to a turn and you're not ready for it and you have to brake and just stop, whereas if you're

anticipating you can see that the path goes sinuously through the trees, and you keep an eye for that. Your whole body is following that and your brain's ahead of the action, working out what's going to come next. So you notice the beginners tend to sort of just stop because they haven't worked out what to do next; they're afraid of going in to a tree or something.

(Kirby)

This biker describes how, when putting in the effort to get to the top of a steep incline, the view suddenly appears, and how different emotion erupts because a visual field that had been limited to twenty yards suddenly opens up, triggering thoughts in his mind, leading to a sense of connection with the landscape. He enjoys the variety of the Commons, describing some parts as bleak, and one part in particular, up above the golf course, left towards Sidmouth, as looking like a savannah (other users of the Common have also remarked on this):

I keep expecting to see elephants and giraffes there because the vegetation at ground level is sort of yellow and you have the isolated trees standing up and it's just like a sort of David Attenborough film. You see that and then you go into this ancient fort, which is only a few hundred yards further on, with oak trees and it's quite different. It's quite bizarre.

(Kirby)

He himself finds the hill fort quite disappointing, particularly after he had been informed of the house that was once there. He is moved by the hill forts in Pembrokeshire but not by Woodbury, and thinks partly that the trees spoil it.

The variety of riding experiences afforded by the Commons is often remarked upon: 'It's a great place the Common. What I like when I take the children there is there's so many different types of terrain; you have small wood coppices to ride in on single mud tracks and downhill places through conifers' (Sam). This variety enables the development of a lot of different riding skills; the Pebblebeds themselves being very challenging:

The other day I thought it would be all right but it actually had been dry, and then got wet and then was really sticky, really slippy. If it's really wet then it slugs you and that's if you ride a mountain bike, it's muddy and slippy. If it's muddy and the mud slides, you kind of know where the bike's going but when it's greasy ... so it's hard underneath and then wet on top, and it's really slippy and that's like riding on ice, so that's quite challenging.

(Sam)

This rider actually finds it quite terrifying to ride over the pebbles but still finds the heathland a 'brilliant, beautiful and special' place. Unlike where she does most of her training (Haldon), where all the trails are constructed from an MTB training perspective, she finds the trails on the Commons to be more natural and the landscape to be wild, in an untamed, non-cultivated sense: 'Yeah, it's not groomed; it doesn't seem to me like someone goes around with a pair of secateurs clipping it all; it's a very natural place' (Sam).

For many, the openness of the Commons is very important: 'When I'm out on the Commons there's a sort of openness; it conjures up images of the Far East to me, it's adventure and even though it's actually quite close (to inhabited areas) it feels remote, you actually feel as if you're a long way from anywhere' (Kimmo). He describes the landscape as giving a sense of being out in the wilderness and says it has a different feel to it than Dartmoor or Exmoor: 'It's not as exposed and in that sort of sense it's more friendly and warming' (Kimmo). But he does not feel the Pebblebeds are wild: 'I know it's looked after by quite a few different groups and interested parties and it's used actively by the military'. He goes on to say that this is a cared-for environment that is looked after in a particular way, which helps maintain its character: 'Because it's such a large space there's a little bit more breathing space, shall we say, for all those different interests – that's the beauty of it'. For him the whole thing is about pleasure rather than a form of transport:

> It's not like you're getting from A to B, that's not to say that some people don't. There are a few people I know who do actually commute across and around that area but more often than not it's more for pleasure than reasons of transport. Going across, riding across the heaths, is very enjoyable because it's one of the few areas where you'll get traffic-free sections of considerable distance; real proper cross-country mountain biking.

(Kimmo)

One cyclist describes how she used to occasionally cycle across the Commons just to get to the other side rather than to enjoy it for itself, but having joined a group she now frequently rides both on and off-road. Significantly, she describes how it depends what one is looking for: 'Are you going up there to keep fit, are you going up there because you want a bit of solace, or, you know, a bit of adventure' (Pamela). She states that it is a place she can be quite proud of and that it is there when she wants it:

> So, if it wasn't there, there'd certainly be something missing I think for the character of this area and your, bit of a cliché, but your kind of quality of life. From, yeah, from physical recreation but also for clearing your head or just sort of making you feel better. It has a lot of qualities; it is very important.
>
> (Pamela)

The last time she had been up there cycling off-road she had arranged to meet with someone but they had got their times mixed up and so she made her own way round the Commons:

> I had a fantastic time as I was discovering paths that I hadn't been on before. I was following other cycle tracks, which I thought might be my friend but they would disappear and you didn't know if you were actually going the right way and I really enjoyed that. I know the Commons reasonably well, you know, but it takes forever to actually get to know it all.
>
> (Pamela)

This particular cyclist feels as if she is part of the landscape when she is on the Common and she enjoys the solitude it can provide: 'There's not many people around, it does swallow people up. Even if there's quite a lot of cars in the car park you don't actually come across that many people except on the main paths'. She enjoys the unexpected variety of the landscape and compares riding under trees, through bracken and over tree roots with wide paths and their vistas of the sea. Stating that she is more of a landscape person than wildlife person, she describes how going on to the heathland lifts her spirits: 'It is stunning and of course it varies from time of year and time of day and conditions'.

Although she believes it is a big enough area to be called wild, she feels the heath is not technically wild because it is a managed landscape:

> But it does have, you know, it does go away in front of you, and it sort of sweeps away and the views are fantastic and rough tracks and gorse and heather and trees and the sort of variety of the

habitats and things and, of course, if you went up there and it was wet and windy it would be wild, you know.

(Pamela)

Reflecting upon whether this is a natural landscape, she states that 'natural' is a good adjective but it cannot be applied to the heathland because it is managed: 'I mean heaths at one time were pretty well self-maintained because, you know, sheep grazing or whatever it was and the commoners, and so it was like part of people's livelihoods and the way of the world that would maintain those areas like that' (Pamela). Thus, although to an extent natural processes once maintained the heath, with human activity sustaining rural livelihoods, that state of affairs has now passed: 'It is more of a managed approach, which will be more regulated, have certain targets and certain aims'. She does feel, nevertheless, that the heathland retains its sense of remoteness and she finds it peaceful: 'Yet it can be a sociable place as well; you see other people doing other things, enjoying themselves' (Pamela).

Asked what it is like riding on the pebbles she says: 'You notice the pebbles when you're on a bike that's for sure because going across the path that goes straight across and down the other side, if you haven't got the right kind of tyres, it's quite a steep bit and you'll start skidding around'. She finds the texture of the pebbles against the vegetation, and the colour of the paths running through it, to present an interesting contrast.

When talking about what could impact on the heathland and how important it is that these sort of spaces are kept she speaks of the campaign to stop a golf course from being built on the Common in the 1970s: 'The battle people put up to save the Common, we should be really grateful to them because otherwise it would be a huge golf course and that would be a dreadful thing' (Pamela). She has noticed an increase in the number of off-road cyclists and is also concerned that there may be too many commercial businesses offering bikes for hire; she hopes that CDE offer a good choice of routes so the load is spread. Like the other cyclists interviewed, she does not wish to see amenities added to the Pebblebed environment.

Relationship with other users

Cyclists discuss their relationship with other users of the Commons:

> I have what you would call the old fashioned type of bell that rings as opposed to bings, and I always, particularly elderly people, I give

them a good ring and quite often, sometimes they say, 'How nice to hear a bicycle bell'. Because if you're cycling along, you're quiet, you come up behind a lady with her dog or whatever and shout, 'Get out of the way' or whatever, they think they're going to be raped or something but if you ring a bicycle bell they immediately know what it is and they just relax and then move out of the way. Actually I always invariably stop or walk round people who are elderly or infirm or whatever. But I don't think all cyclists are well behaved by any means. You see these young lads, the world belongs to them as they whizz past but I think, I mean most of my contemporaries are between fifty and seventy; only a couple of us are over seventy.

(Jack)

He also feels that cyclists do less damage to the terrain than horses do, except when moving uphill when the cycle tends to cut in:

But horses cut the turf and they're heavy on one spot whereas bikers sort of even out and they have a very narrow imprint and move relatively quickly. Well, this article that I read said, scientifically, we do actually less damage to the countryside than the horses do. Which is rather relieving.

(Jack)

Another cyclist feels there should be dedicated cycle paths as this might lessen what he describes as conflict between cyclists and walkers, as well as encouraging more people to ride there. He would not want to create new paths but to install waymark posts to help people to navigate, so they do not get lost or disturb sensitive areas.

In general terms cycling is regarded as a good and sustainable form of mobility in relation to others. However, particularly in the rural context of a conservation area where wildlife may be disturbed by day and night, and in relation to historical monuments, the effects of cycling become more complicated. Potential conflict has involved not just others on the Commons but also how Woodbury Castle, the Iron Age hill fort, is used by visitors. Described as a 'playground' by Bungy Williams, the Senior Commons Warden, Woodbury Castle has been walked on and in for pleasure way beyond living memory. Some erosion had been caused by walkers but this has been exacerbated by some mountain bikers who use the castle to practise jumps and have fun. One cyclist said to us:

> You'll always get people using areas that you're not really supposed to, but where the Pebblebed comes in and the information that's posted around the car parks, that's the opportunity to educate people really. This takes a while and also, through other user groups, it's always easier to put good practice into peer-to-peer networks than other forms really.
>
> (Kimmo)

But some of the bikers ignored the notices put up by CDE asking people not to ride their bikes on this ancient site as much damage was being caused – indeed, during our car park survey we witnessed a large family with their bikes get out of their very large car and then tear off up to the castle's ramparts.

As of 2009, the damage being sustained to the castle was a situation that had to be resolved. If certain members of the MTB public were going to continue ignoring the notices, then measures had to be put in place to prevent cyclists riding on the castle. As part of the long-term management of this scheduled ancient monument, an in-depth archaeological survey of the castle took place in March 2009. This was conducted in partnership with CDE, Devon Archaeology, English Heritage and Natural England, and the geophysical instruments used showed just how deep the erosion was. A plan was drawn up. It was decided that the badly eroded scars had to be repaired, with some of them being formed into steps that would guide visitors up, over and around the Castle. Also, holly bushes and other suitable flora would be planted in order to deter mountain bikers. An archaeological observer was to be present during the construction to record archaeological features that became exposed. Master Builders A. T. Vincent and Sons were contracted to repair the erosion work. The repairs being complete, the steps were built with oak wood and in-filled with gravel from Blackhill Quarry. The site was then re-opened to visitors after fencing was erected around the more fragile scars. Bungy Williams said:

> We just hope that, now the restoration work is complete, people will treat the 'castle' with the respect it deserves, so that it can be enjoyed by the people of Devon for generations to come. The only alternative would be to limit public access, but nobody involved in the restoration effort would want to resort to that.
>
> (Bungy Williams)

Horse riding, co-being and the landscape

Recent research commissioned for the British Horse Society suggests that in the UK over 90% of recreational horse riders are female and more than a third of them are over forty-five years old (Maxwell *et al.* 2012). This is borne out by our research on the Pebblebed heathlands, in which all independent riders seen out on the Commons were women, sometimes accompanied by children. The only time we have observed male horse riders has been during meetings of the East Devon hunt. So this form of recreation is heavily gendered, contrasting with cycling and walking. Another difference is that horse riders, apart from groups of learners riding out from Dalditch stables in the far south of the heathland, usually ride on their own, in pairs or more rarely in larger groups of families or friends. The one established riding school on the heathlands at Dalditch currently has forty-five ponies and horses. Beyond that there are up to eighty independent horse riders using the heathlands on a regular basis. Almost all of them live in villages and farms in the vicinity and ride up to the heathlands and back again. Transporting horses to ride here from further afield, using horseboxes, is not commonplace. So almost all horse riders apart from those using the riding school are local people with local knowledge. Some have been riding here for twenty or thirty years, and in some cases, since they were children. They, in turn, are taking their children out to enjoy the heathlands.

One of the main attractions for all concerned is the absence of road traffic, a major hazard all horse riders have to contend with. Some say that if they did not have this area to ride on and had to use mainly roads they would probably give up horse riding altogether. Both the riding school and independent riders use the whole or large parts of the Commons. The areas they tend to frequent most are strongly related to where they live or stable their horses, to the east, south, west or north of the area.

Those riding regularly may go out on the heathlands three or four times a week for anything up to a couple of hours. Mornings, and evenings during the summer months, are favourite times. Many find riding in the landscape calming and therapeutic, as do some walkers and cyclists. Again this aspect of riding has been noted by the British Horse Society investigation cited above. Horse riding stimulates positive feelings; it may help to counter anxiety and depression and it promotes an appreciation of the landscape with like-minded people (Maxwell *et al.*

Figure 7.5 Group of horse riders on the heathlands

Figure 7.6 Horse-riding partners. Photograph courtesy of Karen Williams

2012: 21ff.). More generally 'equine-assisted therapy' has been widely promoted, stressing the physical and psychological benefits of both riding and being around horses, touching, grooming and taking care of them (Lawrence 1988; Gilbert 2014). Horses and people take care of each other through their mutual interaction.

In this section our primary concern is with horse–human relationships in the heathland landscape as a particular form of embodied relationship that differs substantially from others. While cyclists have their own particular experiential entanglements with their machines, horse riders have a rather different kind of relationship with their animals as sentient beings with an agency and intentionality of their own.

Recent discussions have indicated the importance of understanding horse riding in general and human/animal relationships in particular in terms of a shared sense of co-being and becoming, transcending nature/culture, subject/object, active/passive dualisms in social thought (Argent 2012; Birke 2009; Brandt 2006; DeMello 2012; Haraway 2003; 2008; Hunn 2012; Marvin and McHugh 2014; Maurstad *et al.* 2013; Davis and Maurstad 2016). Horse riders and their horses actively participate in each other's being as part of an embodied relationship that is both physical and mental. In turn this has an intimate relationship to an experience of landscape. If riding horses may contribute to stress relief it also has another purpose for many: as a particular way to experience nature and the landscape. Horse riding provides a perfect motivation for doing so, a reason for getting out and experiencing something Other and different.

Each horse rider, like most cyclists or dog walkers, will have their regular and favourite routes across the landscape. Very few of them can be seen riding across the heathland off the established tracks and paths. They all say they prefer to keep to the tracks both because it is safer underfoot, the vegetation concealing pits and uneven ground, and because there are no adders. Such is the number and variety of potential tracks to follow that riders say there is very little reason for riding off them anywhere else. Should they leave a track it is generally because they have got lost. The horses, like dogs, don't like picking their way through the spiky undergrowth.

Horse riders point out a number of positive benefits and things they appreciate about the heathlands: the relative absence of fences, gates and other restrictions to movement; the absence of traffic; stunning views across and off the heathlands themselves; the diversity of the terrain and the types of areas that one can ride through – high places with fine views, sheltered valleys, open moorland and wooded areas. This provides a varied and interesting riding experience: 'It's so diverse. One minute you can be going down a very dark overhanging track and the next minute you're up Wheathill with clear views of the sea. And the smells up there are absolutely brilliant, the gorse, the rotting vegetation' (Horse rider). They are also keenly aware of seasonal changes in the character of the heathland vegetation and the character of the tracks along which they ride. They develop their own names for tracks based on their characteristics. For two of our informants, Karen and Jackie, who ride together, 'Sandy Gallop' is the name they give to a long, sandy track on a slight incline where the horses can safely run. 'Two-Leg Corner', a sharp bend around which they ride fast, marks an important turning point on a short route out before turning home. 'Jumping Woods' is a short stretch of track where they jump over low piles of logs in woodland. 'Big Circuit' is an extensive ride around the heathlands, the 'Brick Track' is part of the Second World War camp and part of the current RM endurance course, and 'The M1' is a very straight and wide track running to the east from Woodbury Castle across the middle of the heathlands. Riding along these familiar tracks is deeply significant personally because it brings back shared memories of being out together and with their children. The maps drawn by these two horse riders are utterly different from those discussed so far. On the maps produced by the RM, few or no tracks are depicted. By contrast the maps produced by the horse riders are dominated by tracks and car parks are not shown.

Figure 7.7, drawn by Karen, depicts over forty named places and many of these are personal names (e.g. Sandy Gallop, Adder Path, Bluebell Valley, Lollipop Tracks). Numerous tracks are shown crossing the heathland and areas with grazing cattle and ponies. None of these occur on the other maps. Views off the heathland are marked as are boggy areas (very significant for a horse rider). The map covers the southern half of the heathland where Karen rides. The grenade range, quarry, MAFF airfield and Woodbury Castle are shown. The only roads depicted are part of the B3180 and the Woodbury to Yettington road. Little is indicated off the heathland apart from some village names. Unlike the others this is a deeply personal map insofar as it arises almost entirely from a horse rider's perspective, with personal names given to places and particular stretches of the tracks. The second map, drawn by Jackie (Figure 7.8), depicts a similar area of the southern half of the heathlands and is again dominated by tracks and personal names.

The quarry, grenade range and part of the endurance course are indicated together with the positions of bogs and wooded areas.

The nature of the experience of landscape from a horse rider's point of view is very different from that of a walker. They notice different things, in particular the character of the surface over which they are riding. This is absolutely crucial for horse riders. Horse riders, unlike walkers, are much more aware and finely attuned to the manner in which different areas of tracks may change over time on the

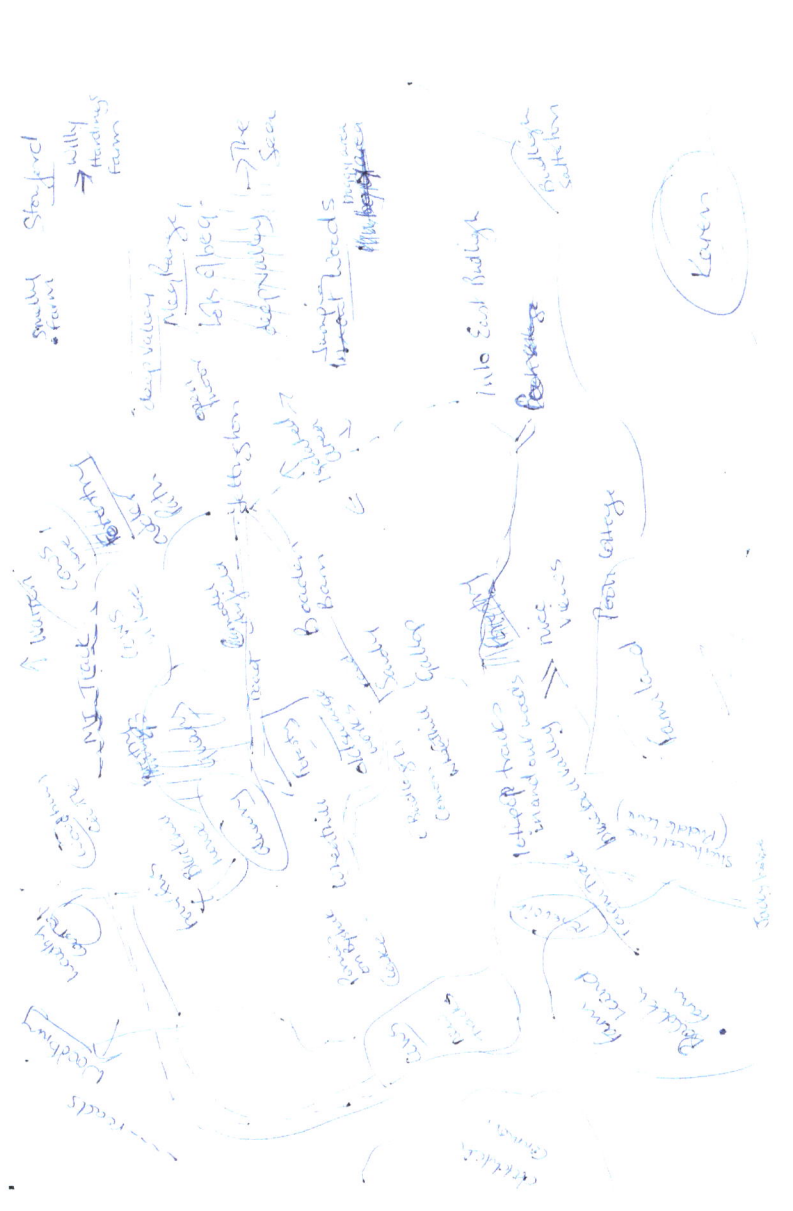

Figure 7.7 Karen's map

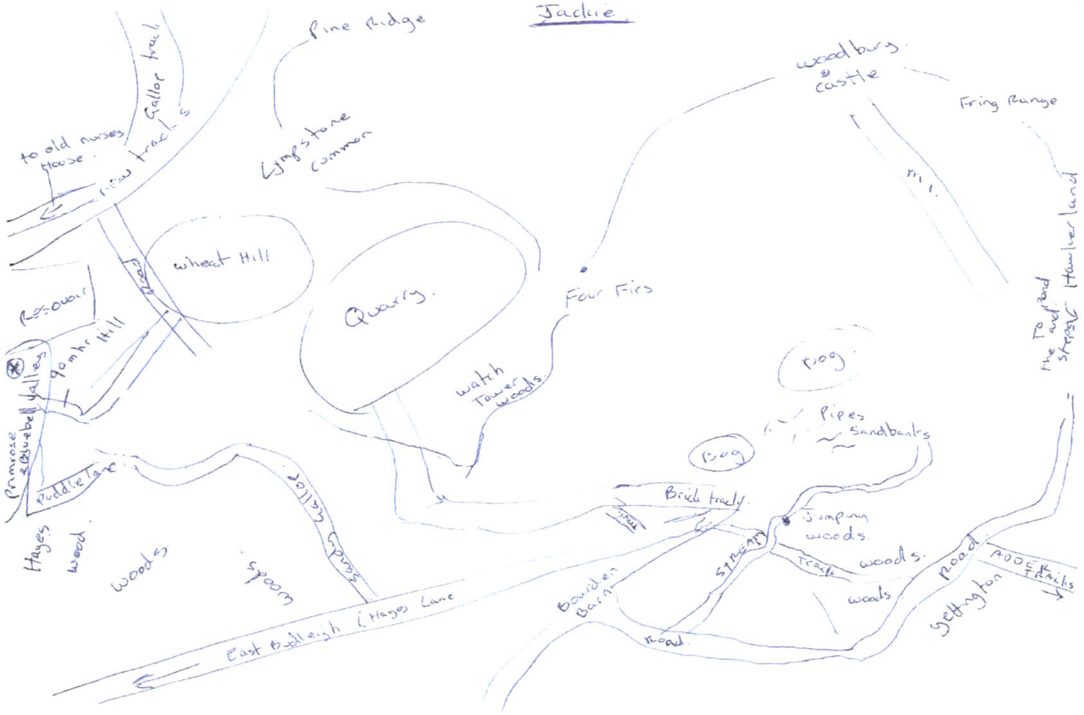

Figure 7.8 Jackie's map

heathlands, in a similar manner to cyclists. They all say that the tracks change. Through time parts of tracks that were once pebbly become sandy, or vice versa. This will make riding along that particular stretch more or less difficult and enjoyable. Sandy areas where the ground is soft are particularly important and exhilarating for horse riders because these are the areas (rather few and far between) that allow you to run or canter. On a horse it is much more difficult going down a steep slope with an uneven surface than going up one, and so particular care and attention to the ground surface is required. Weather may also alter the character of the tracks significantly. After a wet spell horse riders may avoid particular areas because they get so boggy and difficult underfoot.

On a horse you are much higher up than either a walker or a cyclist, and therefore you can see more of the landscape that unfolds before you. Horse riders also tend to go much further than most walkers in the same period of time. So the horse rider obtains an extended view of the landscape in two senses: looking down and across it, and in terms of spatial range. Horse riders say that they see more wildlife sitting on a horse than when out walking: 'You see deer. If you walk you don't seem to. So you see more wildlife on a horse and the horse will detect the deer before you normally' (Horse rider). It is also in some respects less energetic and the experience of the terrain is mediated through the body of the horse: 'You've got the power and the feeling of the horse underneath you and then you've got this beautiful scenery as well and you're not doing all the hard work of the walking you're just being taken through this beautiful, beautiful countryside' (Rider). Steep pebbly slopes make horse riding difficult and challenging in some areas, as do boggy valleys. Sandy tracks without pebbles are areas where it is best to canter because the ground is soft. Favourite areas for most riders are either stretches of track where the ground is soft enough to canter or those where there are wide and extensive views across the landscape, for example, to the west of Woodbury Castle and on parts of East Budleigh Common with views east to the sea. Horse riders orientate themselves in the landscape primarily in terms of familiar landmarks such as Woodbury Castle, like others, but also in terms of the particular characteristics of individual tracks well known to them.

Independent horse riders have noted some significant changes to the landscape: the removal of a pine plantation on Dalditch Common and the restoration of the area to heathland; separately, the huge area swallowed up by the quarry at Black Hill. One mentioned the 'unnecessarily aggressive' fencing now hemming in Dalditch Common restricting access to some tracks she formerly rode along.

Riding on a horse involves a different kind of exercise than walking using different leg muscles to control the horse and move forward. The rider and the horse communicate with each other through the mutual engagement of their sensuous bodies and emotions. For an independent rider this involves intimate knowledge of the horse and its capabilities, because horses may not always cooperate. Any indecision on the part of the rider is likely to be sensed by the horse. In some more difficult areas it is not really possible to look out across the landscape because the rider must instead concentrate on the terrain:

> ... a lot of the time you do have to concentrate and certainly when you are on the stonier places you are watching the stones and on my horse, she's 16 now, and having a bit of joint trouble and I'm very careful where I try to ride her on the Common and I try to pick out the smoothest bits to walk down.
>
> (Horse rider)

The horse itself, as well as the rider, will often know the track if followed many times, knowing where to turn and anticipating a canter or a run at certain points. In winter it may want to take the shortest route home: 'The horses know the routes as well as we do I'm sure. Obviously we make decisions and they don't know which way we are going to go but when you turn down that track they know it will be like this and then we'll have a canter there' (Horse rider).

Horses, however long they may have been ridden by the same rider, can always be unpredictable. Riding always involves a relationship between the person and the horse, their mind and that of the animal: it is a constant dialogue that may be relaxed and harmonious or become tense and difficult (cf. Maurstad *et al.* 2013; Brandt 2006). Riding a horse involves a synaesthetic engagement between their bodies, textures and surfaces, sight and sound and smell. Sometimes the rider may be able to look across the landscape and forget about the animal beneath her in the process of movement. At other times she must concentrate on the relationship with the animal and the area of the track immediately ahead. This relationship also changes with the age and physical capacity of both horse and rider. A 'sparky' horse may calm down with age; older horses will find it more difficult to negotiate steep slopes with an uneven surface and so on. In this way the experience of landscape is always mediated by the relationship established between horse and rider and the duration of that relationship. There is always the danger of falling off and getting seriously injured. Horses can become spooked or frightened by things that take place in the surroundings. Horse riding may be a joy; it can also be a real challenge.

The main potential problems for horse riders on the heathlands involve the other user groups: inconsiderate mountain bikers who flash past and do not slow down, dog walkers who do not control their dogs, model aircraft flyers and the Royal Marines. Some horse riders avoid riding past the model aircraft flying field because their horses can get upset by the high-pitched whining sound of the planes. They will also deviate from their chosen route to avoid areas where they can see the RM are training and engaging in fire-fights. Low-flying RM helicopters sweeping just above the landscape can frighten their horses as do bushes that suddenly start to move (i.e. the RM), causing the horse to stop or alternatively scoot unpredictably up a track: 'You can be going up a track quite happily and the horse suddenly stops and you can't think why and then you look and there is a whole line of camouflaged Marines with guns pointing at you' (Horse rider). However, experienced horse riders are aware of these potential problems of riding on the heathlands and do their best to avoid getting into situations where their animal will be frightened, perhaps asking groups of RM, if encountered, if they are likely to start firing:

> Usually we will see things going on and we will always question them as to whether we are about to ride into a war zone or not because suddenly they can open up and the firing is very loud. The horses cope pretty well with the firing as long as it is not on top of them. But the horses get spooked when they are creeping around in the bushes with their helmets on and hedges growing out of their heads and they suddenly get up and walk and the horses think it's a bush and it suddenly starts moving down the track.
>
> (Horse rider)

Embodiment and landscape

Different ways of riding change the relationship between the rider and the landscape. For example, while walking the rider will typically sit upright on the saddle. Cantering involves being out of the saddle and leaning forward up the horse's neck a bit. This is when the rider will be looking forward rather than from side to side so the faster they move the less, other than the act of riding itself, they tend to see or experience beyond the relationship with the horse. Being on a horse means you are always accompanied in the landscape, never alone. So horse riding always involves negotiation between horse and rider and communication between the two. The two are always in bodily contact in a manner in which a walker and a dog are not, and the movement of the rider is entirely dependent on the horse. This is primarily a matter of kinaesthetics and haptics. The horse rider must, as discussed above, pay attention to the ground surface because the horse has a tendency to slip and the rider has no control over the feet of the horse. On some surfaces they will feel comfortable and may be able to canter, on others they must concentrate solely on the track ahead and move slowly. It is highly significant that all of the horse riders we talked to mentioned the changing character of the surfaces of the tracks that they followed, something that hardly registered with walkers but was also of great significance to cyclists. The rider, unlike the walker or the cyclist, needs to be vigilant and aware of factors that may frighten and upset the horse and avoid them if possible. So getting through the landscape may be a greater achievement and in this sense provide more satisfaction. Riding and controlling the horse requires a different kind of mental activity involving constant and largely non-verbal communication, mainly through the body, with the animal; being one with the animal and ideally in harmony with it and the way that it moves. There is a rhythmic activity using the thighs, calves and legs and the reins to direct the horse forward. There are always two independent minds at work and therefore the experience is completely different from walking or cycling. Horse riding involves a continuous active relationship with another being. The horse rider is always co-present in the landscape with the horse in an encounter that differs fundamentally from those of other user groups on the heathlands.

Conclusions

There are cultures of bicycle and horse riding on the Pebbled heathland that entail an understanding of the environment itself and that involve adapting riding skills according to the various surfaces, inclines, textures and widths of track. Both types of user develop kinaesthetic sensibilities in relation to the terrain and the manner in which they can navigate through it. The relationship of cyclists with their bicycles as a fifth limb, and whether they cycle during the day or night and in what kind of social group, produces a specific sense of space-time and specific evaluations of the landscape, and in a similar manner so does horse riding. Both cycling and riding may be solitary or social but cyclists tend to be more organized and in larger groups. Horse riding is more familial and in our case heavily gendered as female. The bicycle and the horse are both inseparable from human bodily experiences, producing a sense of near/far, up/down, directional coordinates and distant horizons. Knowledge is often tacit, routinized through the medium of the reflective and pre-reflective body, both physical and mental.

The maps produced by our cyclists and horse riders differ substantially and in particular in the absence of roads or car parks for the latter. Named places are fewer and this reflects different kinds of experiences. Cyclists include areas off the heathland itself. Movement for horse riders is generally slower, especially downhill, and usually far more cautious, whereas some cyclists in their pursuit of performance can be reckless. The horse rider has an altogether different kind of relationship with the horse from the cyclist's relationship with the bicycle, as the horse rider is responsible for the horse's well-being. A broken bicycle can be replaced, a horse, with its distinctive individuality and person-like being, cannot. On the other hand horse riders can gallop in sandy stretches of the terrain where cyclists struggle. Far more physical effort is required by the cyclist and for some the effort itself physically taxes the body but may be, for this very reason, exhilarating.

'Challenging' is the word shared by most cyclists when discussing their activities on and around the Pebblebed heathlands. Whilst some of those interviewed cycle on and off-road by themselves, the vast majority rides with at least one other person and often in groups of three or more, especially when riding at night when it appears the focus tends to be on what is in front of the rider rather than around him or her. The sense of vastness and space that is often referred to in daytime riding is lost and the heathland can seem more intimate, which is an interesting contrast to the repeatedly cited 'spooky' characteristic of the heathland at night. From a distance, the cyclists can be quite impressive to watch: their lights and the trails followed are like a moving chain, a lit-up weaving pattern in the night-time. Horse riders only encounter the landscape in daylight and they are generally far more cautious.

Both cycling and horse riding involve shared acts of movement or artistry, but with horses this involves a shared mind and a kinaesthetics linking the rhythms and power of the movements of the horse to that of the rider. The relationship between the cyclist and the cycle is thoroughly mediated by the technology of the machine itself, which may be modified and thought through in various ways to produce different performative acts in moving through the landscape. By contrast a horse rider's relationship with the horse is an intersubjective meeting of minds that, if successful, leads to an understanding between the two and a responsiveness that may heighten the pleasure of both. The emotion a bicyclist may have for his or her machine is a one-sided affair as opposed to the constant negotiation and meeting of minds involved in horse riding. The difference between the cycle as object or thing and the horse as subject and 'person' is fundamental and consequently the emotional entanglement is different. Embodiment is layered in different ways. The bicycle rider experiences through the medium of the bicycle, which itself has no experience or will of its own, but the experience of the horse rider is fundamentally part and parcel of the horse's own experience. Part of this is memory for both human and horse: bad or good rides in bad or good places across the landscape, linked to particular events.

For both cyclists and horse riders there are differing connections with the textured landscape, its curves and changing surfaces. When cycling or riding there is a translation between cyclist and horse rider and landscape, a flow involving the visual and physical, and a making, temporarily, of an imprint by the tyres of the bicycle or the hooves of the horse. This can result in a mobile *being-in-the-world* in a landscape that can change according to terrain, weather conditions and the emotions brought to it by the cyclist or horse rider. Of particular note are the valuable psychosocial aspects of cycling or riding across the heathland, which can, as when walking, become a therapeutic landscape for many.

8
The cry of the Commons: walking through furze

In Britain, walking in the countryside as a popular leisure activity can probably best be described as a recently 'invented' cultural tradition. Until the late eighteenth century walking was regarded mainly as a means of transport used by those not wealthy enough to afford a horse or carriage to convey them on their journeys. For the walkers, journeying was conducted on 'desire paths', the quickest and most navigable routes to the destination. These desire paths would, where permission was given, cross landowners' fields and woods, hugging the landscape's contours.

Influences on the culture of walking for pleasure have been both romantic and scientific. When writing of the first essay written on the pleasures of walking (William Hazlitt's 1821 'On Going a Journey'), Solnit remarks upon how it and the other examples that were to follow over the next 150 years or so were moralizing and inclined to preach on how to walk; Solnit feels this belief in the virtuousness of rural walking persists to this day (Solnit 2001: 118–25). Other influences have come from botanizers such as Richard Buxton (1849), who used long-distance rambling as the basis for naturalistic fieldwork, transforming pedestrian labour into scientific pleasure (Landry 2001: 207). More modern influences include walking as part of the growing 'leisure industry' in post-Second World War Britain (Rubinstein and Speakman 1969), its developing material culture, including what some may regard as peripheral equipment such as pedometers, and the use of mobile phone apps to determine routes. A history of the Ramblers' Association, the Open Spaces Society and other organizations also shows that walking for pleasure has been and still can be political in nature. Although it may be argued that following routes provided by guidebooks or walking leaders reproduces a narrative instigated by others wherein the walking experience is formalized and the opportunities for discoveries lessened, this is perhaps an oversimplification of that which takes place. For example, the 'walkexchange' organization develops educational and creative walks that are free and open to the public. Blake Morris, one of the walkexchange founders who focuses on group walks as an artistic medium, explains: 'We walk to learn about spaces, ideas and each other' (Morris 2015).

Some early writings show recognition of nature's restorative agency, the affordances that may be offered and the way in which emotions, feelings and well-being can be part of an interactive process between the person and their environment. In a poem dedicated to his friend Samuel Coleridge, William Wordsworth wrote: 'From Nature doth emotion come, and moods of calmness equally are Nature's gift' (Wordsworth 1888). In her research on walking in southern England, Kate Cameron-Daum finds that walking in nature does indeed provide her interviewees with a sense of well-being and the opportunity to recover their physiological, psychological and social equilibrium. This perception of well-being is not the passionless, clinical and 'professional' gaze that Foucault describes (Tilley 1990: 296, 310); it is an emotional response and a nurturing of self. It is possible body and self are not seen as such distinct entities. What is objectified is transformed by the self-caring gaze in conjunction with the paths walked. In turn, these pathways and landscapes cannot be seen as set apart from the objectification process but in relation to it. For several of the walkers she interviewed, whatever their state of health, certain paths, locales, and animate forms in nature are embedded in their memory and cultural sense of identity (Cameron-Daum 2008: 35, 39). The medical benefits of walking, both physical and mental, are now acknowledged with a systematic review and meta-analysis by Robertson *et al.* showing walking has a significantly positive effect on the symptoms of depression (Robertson *et al.* 2012) whilst other research indicates that such psychological benefits may be dependent on their social context and an outdoor environment containing greenery for example (Johansson and Hartig: 2011).

Walking as a popular leisure pursuit may be found in other countries besides Britain, including New Zealand (SPARC 2003), where it is the most common leisure activity amongst adults; Norway, where the *Den Norske Turistforening* (DNT, the Norwegian Trekking Association) has created a nationwide network of trails and lodgings, and where Ween and Abram argue that trekking practices encapsulate performances of a ' … banal everyday nationalism' in which nature itself is performed into existence (Ween and Abram 2012: 168), and also France, Finland, Japan and Taiwan (Cristache 2005; Nielsen 2003; Morita *et al.* 2007; Hsiao and Chen 2012).

Anthropological aspects

Walking is one of those taken-for-granted mundane bodily techniques categorized by Mauss as varying depending upon cultural and local conventions. Despite this widely cited definition, until recently, apart from evolutionary theory, it has been largely ignored in anthropology as a topic of interest in itself. A possible reason for this lack of interest may lie with the long-held Cartesian mind/body dualism in which the mind is held in higher esteem than the body. Perhaps reacting against the traditional emphasis on cognition and language in structuralist and post-structuralist approaches, phenomenological studies of the body and its movements as integral to thinking processes have grown in the last twenty years and with them notions of the mindful body (although it could be argued that overtones of dichotomy still remain in this formulation) or person (in a holistic sense) (Scheper-Hughes and Lock 1987). And now walking, together with its possible meanings for the walkers, brings new forms of inquiry and discovery (for example Lee and Ingold 2006; Vergunst 2007; Ingold and Vergunst 2008; Ingold 2010; Lund 2012).

Chris Tilley first introduces walking in his phenomenological perspective of landscape where he shows how essential it is to walk in order spatially to experience, appreciate and understand the possible meaning of the landscape, its locales and the places encountered. Walking can be 'both constrained by place and landscape and constitutive of them' (Tilley 1994: 29). This is a knowledge mediated not by the texts or images produced by others but through personal bodily experiences where, over time, both walking practices and landscape become embodied, yet in writing of these experiences the latter are evoked but cannot be captured (Tilley 2012: 15–16).

Ways of Walking, edited by Tim Ingold and Jo Lee Vergunst, is a multi-disciplinary cross-cultural examination of several diverse practices of walking, distinguishing for example between walking as challenging, as a resource for social and historical knowledge, and as living. An example of the latter is Pernille Gooch's study of transhumance, where feet follow the hooves of the buffalo – a temporal dwelling made fragile and exhausting with the continuing political opposition and attempts to 'wean' the Van Gujjars away from their pastoral life in the Himalayas (Gooch 2008).

In foregrounding the phenomenological importance of landscape and its ties to human response such as narrative, imprint, emotion, the sensual and the importance of textures, these writers provide a sense of how moving forward can be a moving-away-from which is contrasted with the circumambulatory moving on in order to return (Tuck-Po 2008: 32–3). There are openings and closures both metaphorical and in practice, the latter including who is allowed to walk where and how this is transgressed in order to survive (Widlock 2008: 58–61).

A temporary dwelling

When listening to the heathland walkers relate what they experience and feel when walking, it becomes clear that Maxine Sheets-Johnstone's proposal that movement and thinking are not separate happenings is valid: they are 'aspects of a kinetic bodily *logos* attuned to an evolving dynamic situation' (Sheets-Johnstone 1999: 489). When relaxing or reflecting, movement can enable or mediate the thoughts and feelings of the walker; the environment walked in is also likely to be part of this thinking and feeling process. The heathland is not an empty space, not an appurtenance, but a world that is constantly folding, unfolding and refolding. It is an environment in which a social and cultural response is embodied. On occasion it may, in Deleuze's terminology, feel interiorized, wrapped 'in an instance that can ultimately be called "personal"' (Deleuze 2006: 144). The anthropologist Tim Ingold speaks of what he calls a 'dwelling perspective', which is an alternative to the nature–culture dichotomy (Ingold 2000). Thus walking on the Common can perhaps be regarded as a temporary dwelling, an interaction and engagement with a landscape that changes according to season and weather and to the walkers' perceptions, emotions, interests and memories. This we explore in detail in the rest of this chapter.

The cry of the commons: motivations for walking

> They've got this weird melancholy cry that you can hear from miles away.
>
> (John, walker)

A fascination with buzzards is just one example of the phenomenological nature of the walkers' relationship with the Pebblebed heathland; the majority of the walkers refer to the heath as 'The common', an umbrella term for all the heathland commons. All the walkers we spoke with cite sensory and emotional factors that help build this relationship and these include the smell of gorse, the feeling of solitude, the landscape's wildness and the beauty of being in its open space.

The walkers' main motivations for walking on the Common are often interlinked. They include well-being and how this is felt through movement and the use of the senses; a love of nature; pleasure in being outside either as a solitary activity, with a friend, or while participating in a led walk. There are diverse ways in which the walkers use the heathland footpaths to take them in to spaces where they are with nature. For some, it is a close and intimate relationship; for others there appears to be a need for space, if not distance, between what they perceive and themselves, and these particularly favour a vista that involves sight of the sea. The walkers have very different reasons for being in nature. It may be curiosity or knowledge-seeking, or it may be more to do with the release of anxieties.

The walkers

John is the source of our opening quote above. In his childhood he lived surrounded by acres of rhododendrons. When he and his wife retired they moved to Budleigh Salterton, which – as well as being by the sea – possesses the acidic soil these ericaceous plants require: 'I'm trying to create a little paradise' (John). Although recent arrivals, John and his wife have known the area for ten years as they came here for holidays. Dalditch Common is the common they know best but he is also particularly interested in history and is acquiring knowledge of the military history of the commons. Other walkers also moved to the area after retiring or being made redundant and several joined the Otter Valley Association (OVA), often becoming walk leaders.

Another walking couple, June and Richard, were born in the area over seventy years ago and have continued to live here throughout their lives. They are certainly amongst those with the strongest memories of using the Commons in their childhood and remark upon the freedom there was then: 'You could go anywhere and do anything without fear of any problems. It's not quite the same nowadays.'

Margaret used to watch the swaling when she was a child and her father, Alan Toyne, was the key figure in the campaign against the building of a golf course on the heathland (1971 to 1974). She remarks, 'It was a huge issue for Devon. It was one of the first big campaigns to save a piece of land'. She is naturally immensely proud of her father, particularly in view of the diminishment of lowland heathland throughout the world. Benjamin, who was born in this area and returned once he had retired in 1999, tells us that his predilection as a child was not for the beach but to travel up to Squabmoor, and interestingly he feels that there has been no change in landscape terms to this particular area of the heathland.

Susan has lived in Lympstone for over twelve years. Although she enjoys walking on her own, she is also a member of walking groups such as the Exeter Outdoor Group. Another walker moved to the Pebblebed heathland area over thirty years ago with his wife. A keen walker, he organized the publicity for the archaeological Pebblebed Project and its open days. The project has changed his life in that what were once leisurely walks in favourite green lanes with their trees and flowers have now become more focused on the pebbles themselves.

Three of our walkers came to the area having either been born in or lived in Africa. One of them knows the Common very well having lived in East Devon for the last thirty-five years, twenty of which were spent working for CDE. The other two walkers told us they choose to walk in quiet areas because they have two autistic sons: Woodbury Common is their favourite. The location is not all that is important: 'It depends on where the wind is, what the weather's like and how you're feeling', says Alice. She and her family particularly like Woodbury because it reminds them of the landscape they once walked in Africa: 'It has a Nevada effect, which we had in South Africa. There are just certain areas in which you stop for a moment and think "Wow", and if you catch it with the sunset it's really beautiful'.

The physicality of walking

The physicality of walking is not such a mediated experience compared to the mountain bikers and their bikes, and the horse riders with their horses. Although the Marines also use their feet this is more likely to be when undergoing a nine-mile speed march, and this is not comparable to the pastime of walking for pleasure with its intimate sense of engagement and equanimity. One walker, Paul, enjoys walking on the heath in his bare feet: 'I've got quite hard feet and I can walk on the pebbles. I've been for many cold walks up there in bare feet'. Walking on the heath involves movement, pauses and on occasion, rhythm, the latter particularly when the walker is intent on reaching a destination at a specific time. It also encompasses emotion, feelings and thoughts, as well as memory and comfort. Although there are flat, easily walked routes, others are less accommodating and some walkers find them quite challenging.

The character of the heathland landscape

How individuals feel about the character of the heathland obviously varies from walker to walker. Indeed it may vary according to which area is being walked in, the views afforded and the conditions of the terrain, which is frequently affected by the weather, but it may also depend on the purpose, if any, of the walk and – as Susan states – one's personal mood at the time of moving in this landscape. John says it reminds him of Hardy's Egdon Heath and notes that although he finds it does not contain the slightly menacing air described in *Return of the Native*, he feels the Pebblebed heath is its own master and should be treated with respect. Pauline has felt fear, stating that the heath can be frightening and threatening at times. Allison does not find much of the heath an intimidating place but does describe the fir plantations as being like an Eastern European fairy-tale forest: 'You know, it looks like a deep, dark, Polish forest with wolves …' (Allison). Some walkers describe the Common as being beautiful, others find it fascinating and inspiring, and several walkers remark upon its openness and the sense of freedom one gets from walking there. Many walkers speak of the changing colour of the scenery over the seasons – greys, lilacs, yellows and soft greens – and reflect upon the diverse range of materials that lie beneath one's feet – pebbles, sand, grass, mud …

Engagement with the heathland

Two of the walkers we spoke with (Sylvia and Peter) walk from their home two miles up to Woodbury Castle every year on Boxing Day to join the East Devon Hunt (EDH). It has become a tradition, having been held here for over 120 years often with between sixty and seventy horses and three hundred foot followers. Since the Hunting Act of 2004 the EDH meets for trail hunts, with dogs tracking a trail laid with a scented piece of cloth. Not everyone is happy with the trail hunt, including these walkers: 'It's a nice event to see and we don't get steamed up about hunting foxes. I would be perfectly happy for them to carry on hunting foxes, which are vermin in the countryside but it's been stopped so they're drag hunting instead and good luck to them' (Sylvia, walker). Her partner remarks there is plenty of support for this across the social scale locally and speaks of how coachloads of people used to arrive from neighbouring towns such as Sidmouth with buffet lunches laid out at the Castle:

> The Clinton Devon Estate has very rightly closed off a lot of the car parks so now coaches can only go into one car park, which, I think, has stopped them coming so much because people who come on the coaches don't want to walk; they want to be delivered to where the hunt is.
>
> (Peter, walker)

Although Peter now walks there far less often than he used to, Sylvia visits more often; until recently she took the Girl Guides there on field trips as it is a 'graspable size'. She remarks on how there are always flowers to see and identify, and birds such as stonechats and nightjars to listen to. Having been brought up near the lowland heath of Sutton Coldfield, which also has pebbles and where she developed her knowledge of wild flowers, she loves the heathland: 'So this in a way is coming home for me because it's exactly the same flora and fauna as I was used to as a child' (Sylvia, walker).

Edward brings his family on a regular basis throughout the year. He says he loves being out in the open air and wishes to instil this appreciation in his children. He feels the Common provides a valuable education for them in a number of ways but particularly for its wildlife and history: 'For me and obviously for the family as a whole it's very important'. He tells us they have great fun with the children and the pebbles:

> We try to find the smoothest, the widest, the knobbliest pebbles. Then we'll get the kids to find a red one and a white one and it gives them an extra focus. Everything is a learning experience and if you can make it interesting as well then they're more likely to listen and learn.
>
> (Edward, walker)

Led walks

Andrew, a walk leader, says there are so many paths it takes some time to get to know your way around. It took him a few years to get to know the area before he was confident enough to lead a walk. He likes the openness of the Commons but not the fir tree plantations: 'They are certainly not so interesting and they don't support any sort of plant or bird life'. He prefers to walk in the wintertime and on one January walk it had been snowing: 'It was really magical, the Common looked entirely different'.

All of the walk leaders agree that it is nice to get people who have a particular expertise on these led walks as they point out interesting things that they might otherwise have missed. Ian leads the OVA walks three or four times a year, some of his walks going round by the hill fort: 'I always try to imagine what it would have looked like. It must have been a formidable construction at the time'. Duncan believes that the led walks are particularly good for single older men and women as it provides them with an event that they know is going to be sociable, and Andrew feels led walks fulfil a need, especially for single women who may not want to walk alone. He actually prefers to walk with just his wife and have the place to themselves: 'We walk in silence and find that on a lot of these led walks people chatter away and don't really look where they're going. You get back and they say, "Where've we been?"'.

However, walking on one's own without a leader creates the challenge of navigation. Like other walkers, Susan has found that the pathways are not very well covered on the maps and thinks this maybe because the Marines have developed their own. To become a walk leader she needed to learn how to navigate open land: 'Even where it is open access land it's fairly overgrown and I thought to hell with this business of sticking to footpaths, I'm going to go and walk towards a cairn or something'. But her attempt to create her own pathway failed: 'I got desperately caught up in brambles. I mean I really came out absolutely streaming with scratches on my legs. It was quite frightening'.

Several walkers appreciate the information boards and labelling of crop fields but Edward disagrees and says if he wants to know more he does his research beforehand, or after the walk: 'Having the information boards at the car park is great. Once you start putting them in the open spaces where people can see what they're looking at from various angles, then you impose on what people can or can't see; you impose on the feeling about the place'. Although the Common is not common land Sylvia remarks that this does not actually affect the way people feel about it or use it: 'In fact, it's probably better looked after because of the Pebblebed Trust, better cared for than if it had been just common land'.

Change

Ian remarks on the remaining traces of the past:

> It's always interesting to imagine that the Bronze Age and Iron Age people might have tramped the same paths that we tramp today. I sometimes wonder about the links between Woodbury Castle and neighbouring hill forts and what routes they would have used to get from one to another; whether they ever travelled by boat.
>
> (Ian, walker)

Whilst at Woodbury Castle we spoke with a geologist who has been visiting the Commons for over thirty years. He finds the changes at the hill fort are very small: 'It's a very resilient structure'. He is also pleased with the new steps that have been put in to deter mountain biking. Over the years he has brought many children from the Woodbury Saturday Club to visit the Commons, with the hill fort providing a wonderful environment in which to play hide-and-seek.

Benjamin is pleased with the Second World War encampment returning to nature, but Margaret is not happy with the changes to the Common since she visited as a child, particularly since 2005:

> They've done a lot of management. They've cut huge swathes of gorse, closed many of the car parks and inside the hill fort it's now a big flat space. It was very overgrown for a long time and not really managed at all, which I much preferred, because it now looks so pristine and it doesn't look and feel its age.
>
> (Margaret, walker)

Paul describes the heathland as having become gentrified and that there is a build-up of biomass that never used to be there: 'So it has changed. Even the habitat has changed'. He has quite strong views on scraping as a method of management:

> It's well recognized that to keep it as a heathland you have to burn it. Scraping is no answer at all. It's not a natural way of doing it and it's not how it's been done for the last three thousand years. They're going to completely change the whole nature of it by scraping it. Burn the bloody stuff, that's why it is what it is. Don't be so bloody precious, you know.
>
> (Paul, walker)

Natural wildness?

Creating a mosaic that gives the heathland a natural appearance and is of maximum potential for its wildlife is very important to the CDE. The concept of what is 'natural' provokes response from many of the walkers and several feel the Common is both wild and natural, despite knowing how much management goes in to maintaining it. Benjamin describes the Commons as 'untamed' and finds the sharp divisions between the formal agriculture and the scrubland Common very interesting: 'You can get a perception of how things over the years could change'. But Duncan feels quite hemmed in because of the way in which the Common is dissected by various roads, adding that it is noisy and too close to civilization to be described as 'wild'. Yet Ian describes going to visit the Common on a June evening, listening to nightjars and snipe, and says you do feel in a very wild place.

Memories

The 1971 announcement that the CDE was proposing the construction of two golf courses, one on Woodbury Common and the other on Colaton Raleigh Common, was met with great anger by many of those who visited these Commons at that time. Walker and local historian Sally and her husband, Ramsay, were amongst those joining the rallies organized to oppose it. Thousands of signatures against the proposal were collected within days. Sally, who has done much campaigning against developments such as these, says, 'Oh, looking back it was so heartening. It started a terrific outpouring of opposition and being part of an audience that felt the same way is a nice change'. She describes their public meeting at Woodbury Village Hall as being packed with people, the room too small to accommodate everyone:

> They crowded the hall, stood at the back, stood in the porch outside; some people to have their say put their heads in through open windows to address the meeting. Loudhailer equipment relayed everything that was said to those outside and couldn't get in to the hall. It was exciting, fascinating and, you know, feelings were running really high and in the right way as far as I was concerned.
>
> (Sally, walker)

Sally describes the strength of feeling about the Common that existed then as being very much an emotional attachment.

Although it is the vision and perspective of the CDE and RSPB wardens that is responsible for the heathlands we see today, it is the work of the campaign against the golf course that provided the opportunity for this vision to succeed. If the golf course had gone ahead, despite limited SSSI protection given in 1952, it is doubtful that the heathland would have been awarded the substantial protection it has now, and for this we are indebted to the passion of Sally and her fellow campaigners. Their human intervention and protection of the heathland has enabled its conservation. Ironically, Sally speaks of how, for some people, nature is tidied up and becomes a status symbol; in fact it may also be argued that in seeking to preserve or re-create the Pebblebed landscape which 'houses' certain flora and fauna that have existed here for hundreds of years, conservationists are creating or maintaining arranged, tidied and perfect places. Parks in our cities can be seen as examples of perfect, organized space; does conservation of the heathland contain elements of such organization? Perhaps, but for Sally this is not an organized space but a valued place which bears traits of a Foucauldian heterotopia, in which fear of loss of biodiversity may be linked to notions of heritage and a place of time that Foucault describes as being in itself 'outside of time and inaccessible to its ravages'(Foucault 1986 [1984/1967]: 26). Sally's memories of the successful campaign against one golf course's construction on the Common are uplifting and part of this phenomenological place of time. So it is positive to view the surveys she has and is conducting – her photographs and maps will, as she has said, encourage others to view for themselves the beauty of this countryside and in so doing this can become 'a familiar part of our inner landscape of "home", a cultural link to our ancestors down the ages' (Sally; see Elliot and Wickenden (n.d.)).

For Benjamin the Common is sometimes a place to re-trace the paths he used to use: 'I can pretty much find my way around the Squabmoor area without guidance and still on pathways I used as a child'. He describes himself as having a sort of mental map of how to walk round there. He says that he has always felt the heathland to be a part of his life: 'You know, it's one of those things that sticks to you'. When he was in his forties Paul enjoyed walking the Common at night. He found that it was not wise to move off the main path, and once saw a moonbow there: 'There was this grey ghostly rainbow, all in silvers and greys. Apparently they're very rare and it was an amazing thing to see but you'd never get that if you didn't walk out at night; you probably wouldn't notice it in a car because of all the light pollution'. He tells us that walking the heathland at night 'binds you into it'; you have a relationship with it.

Sylvia's parents, when on holiday in Sidmouth before they got married, had actually got engaged on Woodbury Common: 'They had this photograph of one of the clumps of trees, which has now gone on the Common. It was always in our home, and the house I was born in was called "Woodbury"' (Sylvia, walker).

Mark says he has little knowledge of the archaeological features on the Commons but that when he was a boy he dug up an Iron Age artefact at Woodbury Castle. He and his mother took it to Exeter Museum who told them that it was a digger of iron-bearing sandstone, usually found only in the Netherlands. He still has it fifty-eight years later and it sits on a windowsill in his home. In the early 1960s Ian learnt to drive on the old concrete roads that had been laid down in the Common during the Second World War. He speaks of the time when, in the winter of 1962–3, Squabmoor reservoir froze over and they went skating. Richard recalls that when there was a bad drought in 1976 people came up to the Common to cut ferns for animal bedding. Also, when Blackhill was a working quarry, being tenants of the CDE meant they were entitled to free hardcore from the quarry to make their roadways and pathways. Benjamin remembers visiting what was left of the Second World War encampment with his mother when he was a boy:

> The whole place was littered with little square lots of two or three courses of brick and a chimney, and a fireplace – magnificent. And I remember my mother was very envious as she always thought these fireplaces were much better than the ones she had at home, and if only she could take ones of these home it would be great.
> (Benjamin, walker)

The walkers' personal historic links and memories create a sense of ownership. Sylvia says she does feel a sense of ownership for the heathland and she is not alone in feeling this. Richard says: 'Having lived here all my life I've always been associated with it so you are inclined to think it's sort of "ours"'. And 'our' Common is exactly the phrase used by Margaret when she speaks of how they, the anti-golf course campaigners, were trying to save 'our piece of land'.

Contestation

As discussed above, the Pebblebed heath has many other uses besides being a place to walk. Of RM activity Duncan says: 'They pop up sometimes but you just accept that they're there'. Andrew too has no strong objection to the Marines using the Commons: 'They're a bit frightening when you come across some of them, and some of the water holes they have there as part of their challenge can be a bit daunting when you're walking and come across these very deep gullies'. Chuckling, Peter says: 'Marines leaping out or hiding in bushes when you're going for a walk, or you suddenly feel or hear the bush talk, is an added frisson to walking on the Commons'.

Andrew says he finds the model aircraft flying a bit intrusive when he walks on the Common for peace and quiet and hears the whining of the planes. He also regards them as a perpetual fire hazard. Ian finds the planes a bit noisy too but he thinks that on the whole, people use the Common very responsibly. He also states you occasionally find cyclists on footpaths where they strictly shouldn't be but that it is a matter of live and let live. And Mark is not particularly enamoured with the MAFF activity either, as when he was about eighteen years old, a model aircraft bounced off the roof of his car and badly scratched it.

Dogs are the focus of complaint by some of the walkers. Peter says the dog walkers are alright if they clear the mess up, and his wife, Sylvia, comments that according to the notices in the car parks, dogs are supposed to be kept on a lead during the bird nesting season: '*Nobody* but *nobody* keeps a dog on a lead on the Common'. She feels that many dog walkers do not control their dogs properly when out on the Common and that this is a drawback for people who do not like dogs. Liz, a walker who has had much experience of working with organizations connected with the countryside, including Countryside Stewardship and working on a website that is to do with conservation, recreation and conflict resolution, says she does grumble sometimes about the dogs and the mess that is left: 'I like dogs, you know, but it's their owners and when walking I have occasionally been chased or yapped at or nipped by unruly dogs … but most people walking dogs up there are lovely, and the dogs are lovely and well trained'.

Perspective of dog walkers

We have observed that the most frequent users of the heathlands are dog walkers but the majority park at Woodbury Castle, follow a quick circuit, and then leave. As can be seen, other walkers often mention dogs and their owners, sometimes voicing complaints about mess that is not cleared up or dogs that are unruly, and so it is interesting to listen to the views of two walkers who visit regularly with their dogs and for whom the experience is more than toileting their dogs.

One dog walker is also one of our artists, Caroline (see Chapter 9), and it is evident that the walk itself has two important purposes. One is to exercise her dogs, the other is the way in which a quiet walk allows her to reflect upon her work in progress. She has favourite places to walk in for an hour or so, and at the time we spoke she often walked with her two dogs a little further on from Squabmoor Reservoir, choosing to walk here as it was then most convenient:

> I can do a small circuit with the oldest dog, put her in the car, and then do a slightly bigger second one with the younger dog. I do a sort of double one and that's a nice walk when the weather's bad because it's through quite a lot of trees and it's undercover so you can be sheltered.
>
> (Caroline, dog walker)

Caroline says that her dogs make the difference in going there and that although she enjoys walking she probably would not go so frequently. Poignantly she says: 'Mmm, I can't imagine not going there but it will be hard, especially once I've not got the dogs because I shall miss the dogs even more by doing it'.

Chris frequently walks the heathland with his dog, Tor. He too often follows an established route and says that this is true for many dog walkers on the heathland; the walk is one that they have become familiar with, without the stress of having to plan a route or think about the walk. He describes what is initially a regular walk from Hawkerland, up Robin's Lane, onto the open heathland. He remarks on the butterflies and spring blooms that can be seen alongside the lane. He also comments upon the way in which Tor loiters behind him, sniffing and leaving scent marks. Soon, Tor is no longer behind him but in front, always in his vision, whether main or peripheral. He says that once up on the heathland everything changes:

> Here there are no flowers and few butterflies. Particularly at this time of year (early spring) the heath seems utterly dead, lifeless and colourless: yellow-grey, white and brown. The only colour is the bright yellow of the tall-growing European gorse. This contrasts with the verdant green of the fields beyond where everything is bursting into life. Up on the top of the hill I can now look out across the landscape. Rather than looking over the heathland I tend

> always to look off it, across to the fields below. I sit down. Tor lies down a short distance away. The heath becomes a vantage point for me to experience that which is beyond. It is the relationship between heath/non-heath that is all-important.
>
> (Chris, dog walker)

Spotting a new firebreak he decides to follow it and see where it leads; the familiar is now becoming unfamiliar and he says Tor no longer dawdles but is running up ahead:

> Over a local summit we drop down to a shallow coombe in which a delightful little artificial pond has been made … complete with ragged bulrushes. Tor runs to the pond, throws herself down in it, and starts to drink the water. Hating the gorse she rarely deviates from tracks across the Commons. She is a track follower and will always wait at the junction of tracks to see which one I decide to take … When I get home I notice how numerous and loud the birdsong is in the garden compared with the low and muted calls experienced on the heaths. A different environment and a different species.
>
> (Chris, dog walker)

For both of these walkers, on many occasions their dogs are the main purpose for walking the heathland and consequently they partly experience the heath via the medium of the moving dog, as much attention being paid to the dogs as to the landscape that is being walked in. Dog and dog walker are co-present in the landscape but in a very different manner from the relationship between horse and rider discussed in Chapter 7 as both move independently but in relation to each other. The most important aspect of this is that the dog walker in part experiences the landscape through the medium of his or her dog, his or her movements and deviations. Little of the heath's wildlife is seen and this may be to do with the presence of their dogs, but as the dogs themselves rarely leave the tracks they do not disturb nesting birds or Dartford warblers. Chris has also noticed how the sensory experience of the heath is limited for both walker and dog, his dog finding the lane leading up to the heath as being of more interest. Caroline is drawn to the comparative shelter and quietness of Squabmoor as an environment in which she can think about her artwork whilst exercising her dogs. But, it is the heathland's open areas that are of importance to Chris: 'The view is paramount'.

We have also not observed many women walking alone on the heathland, particularly without dogs, and this is reflected in some of the interviews. Sylvia says she would not walk on the Common by herself, not because she is afraid of being by herself but just in case:

> I have to face up to how old I am now and if I did have a fall or something, and there have been cases of people having a fall and lying out there all night. It's much pleasanter to walk with someone else but I've got several friends who'll walk up there with a dog, on their own.
>
> (Caroline, dog walker)

Susan is slightly afraid of strangers and strange men when she is on her own there, possibly more so than in other places, such as the coastal path or Dartmoor, but she is uncertain why this is the case: 'Maybe it comes from a childhood fear of commons and witches, you know, a lot of that fairytale kind of stuff'. Another female walker who was put off walking alone on the heath is Janet. Prior to moving to Woodbury she lived in Farnham, and Frensham Common nearby had a bad reputation for assaults on women. She finds the associations just too strong. Elinor also says she doesn't know whether she would actually want to walk the heathland on her own: 'I find it's a bit too isolated and I wouldn't want to be there by myself, but then that's as a woman speaking'. In fact, one of the Marines we spoke with said he would not want his wife or children to be on the Common by themselves either.

Eight of our walkers produced sketch maps for us. They vary quite significantly. One shows most of the heath, the remainder show those areas with which the walkers are most familiar. Four show a particular sequential walk across and around the heath. These all mark car parks – the beginning and end of the walk – and topographic features and places encountered along the way.

Climbs, muddy places and bogs, woods and pools are shown. Roads are almost absent. The features depicted on all the maps are either highly localized places or distant views off the heath to the coast and surrounding hills, a combination of the intimate and the remote in which the landscape seen from the heathland is as significant as the heath itself. Three of the maps mark wildlife sightings, the rest do not. Two show Woodbury Castle and one marks the model aircraft field and two areas used by the RM for training. The maps contrast significantly with those produced by others such as horse riders, cyclists, and environmentalists, providing additional insight into those features of the heath that are most significant in relation to how it is encountered and used.

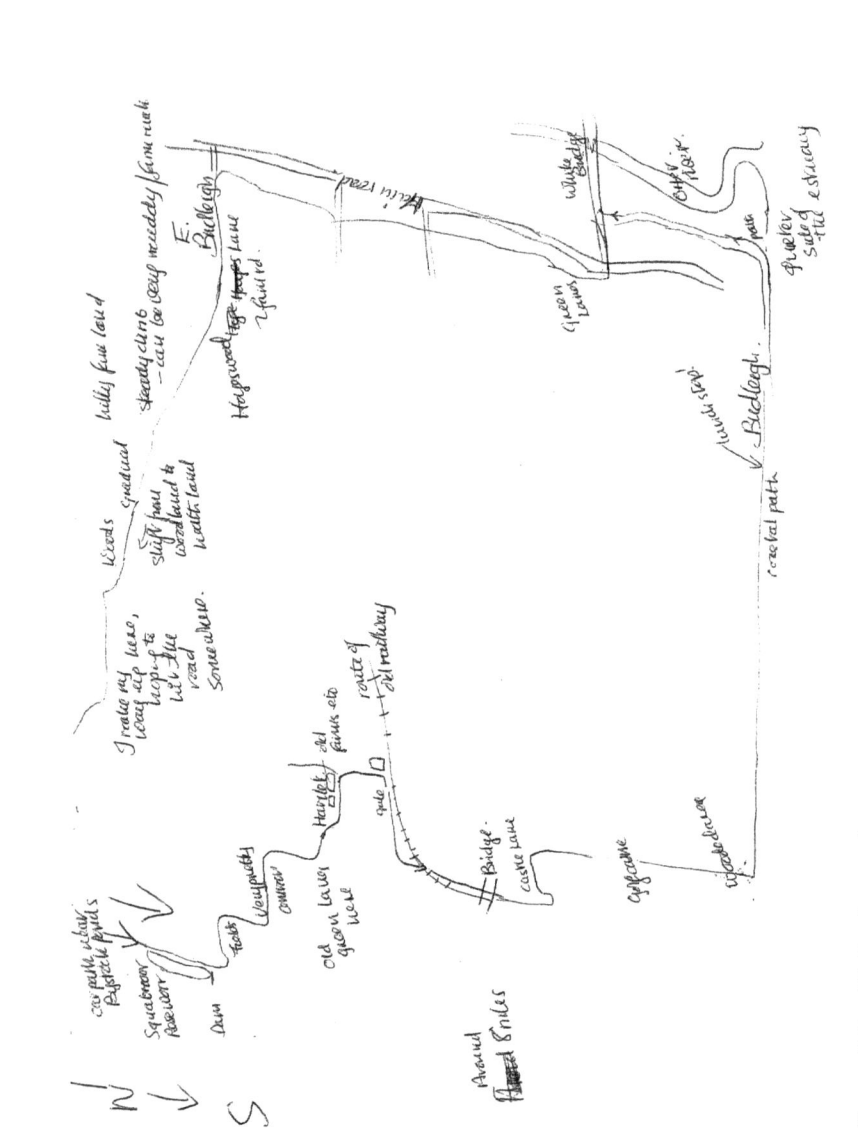

Figure 8.1 Walker's map 1

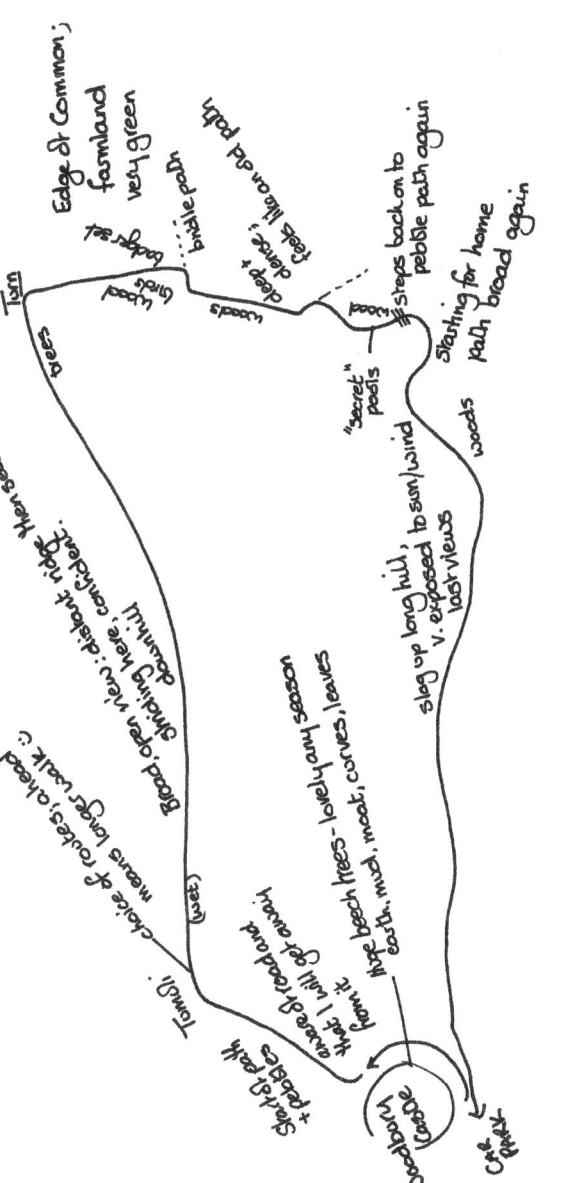

This walk is about 4½ miles long. The muddiest part is near the beginning. I only take this route when I have plenty of time – satisfying itself! There is a lot of variety of both views and terrain.

Figure 8.2 Walker's map 2

Conclusions

Spatial relations within the heathland are complex. Hebdige writes of how space has come to be refigured as 'inhabited and heterogeneous, as a moving cluster of points of intersection ... which cannot be reduced to a unified plane or organised into a single narrative' (Hebdige 1990: vi–ii). The heathland and the ways in which movement within it is experienced by walkers and others are not homogeneous; this too is not a single narrative. Rapport describes walking as involving both a phenomenological objectivity – 'a common human embodiment' and a phenomenological subjectivity – 'an individual consciousness engaging in imaginative projects of disembodiment and otherness' (Rapport 2008: 32). When we listen to the heathland walkers we can see how important the heathland, its paths, scenery, views, wildlife and history are to them. For many it is a therapeutic landscape that promotes physical, mental and emotional well-being, a place travelled to either on foot or by car, away from the pressures of everyday life. And the biographies of the walkers have given us some indication as to why they choose to walk on the Common; the proximity of the Common to where the walkers live appearing to be the common factor. But walking here can mean different things to different people at different times in their lives. Here, Rapport and Dawson's (1998) important contribution to the discussion regarding how identity may be configured through temporal and spatial relations is very apt. To note that movement has become fundamental to modern identity and its experience is an essential component of everyday existence (Rapport and Dawson 1998: 4–6) does not necessarily relate just to the globalized world of movement that they discuss. For those returning to walk on the Common in their later years, it becomes clear that the physicality of movement, the imprint both of foot and memory on the Pebblebed heathland, is of much importance. Yet the general recognition that 'not only can one be at home in movement, but that movement can be one's very home' (Rapport and Dawson 1998: 27) resonates well with the heathland encounters of our informants. For many of our walkers the heathland has become another kind of home and it clearly affords a different kind of dwelling to the home and the built environment. In moving through and encountering the heathland in its different moods and seasons they create a new sense of the self through the medium of walking in this landscape.

The heathland meets the different needs and intentionalities of all the walkers with its environmental affordances: challenging, restorative, an opportunity to gain knowledge about the wildlife that dwell on the heath, its diversity is appreciated. A journey through this landscape can thus be transformative. Although some people are nervous of walking alone on the Common, the Otter Valley Association and Exeter Outdoor Group (Exeter Outdoor Group, n.d.) offer the chance to explore this environment in their led walks, which are often educational. Other walkers prefer their walk to be a less social activity; they may be intent on observing nature or finding the quietness and solitude they seek on the heathland. Walkers and heathland meet and act upon one another, the embodied agency of the landscape, the sensed and sensing body; each is produced in mutual, mediated relation, through movement, consciousness, perception and emotion.

9
Art in and from the landscape

In this chapter we provide a brief review of relationships between artists and landscapes, considering environmental art and the inspiration provided by walking in the landscape. We finally mention developing relationships between the practice of art and the work of anthropologists to provide a background to our own consideration of the arts in relation to the heathland landscape and in relation to a collaboration with one particular artist, Priscilla Trenchard.

Wittgenstein claimed, 'what *can* be shown *cannot* be said' (1981: 50). Yet in his lucid account of 'circuits of subjectivity', in which he critically examines, amongst other aspects, the sometimes ambiguous but essential relationship between words, gestures, memory and art objects within the context of oral history work, Cándida-Smith writes of how we appreciate visual and performing arts because they provide us with aspects of experience and feeling that elude words: 'but then we must use words to share processes that communicate on other levels' (Cándida-Smith 2003: 8). In interviewing the artists whose works involve the heathland we are concerned, therefore, with the essence of the work itself, how it is produced and from what embodied intentionality and inspiration it may have been assembled – the unfolding relations between the agency of the landscape, that which the artist brings to the landscape, and the agency of the artwork itself. And to frame the discussion we turn to the words of the poet T. S. Eliot, when he writes of Andrew Marvell's verse 'making the familiar strange, and the strange familiar' (1950: 259).

Dufrenne considered the manner in which an inspired artist may conceive of the process of producing the artwork as a being that he must promote: 'a being-for-the-artist, a being which is anterior to his act' (Dufrenne 1973: 33). In Gell's Maussian reading art objects may be considered as persons (Gell 1998: 9) or 'distributed persons' (Gell 1998: 103). Küchler comments that from this perspective both an artwork and the inspiration 'of intuitive associative thinking' are a locus of agency in the unfolding of person–object relations (2006: 86). We now consider this process in relation to landscape art.

Fragile environments: nature and culture

As we have seen the Pebblebed heathland is a fragile environment that requires constant nurture. Author and environmental activist Berry writes, 'a culture that does not measure itself by nature, by an understanding of its debts to nature, becomes destructive of nature and thus of itself ... The only thing we have to preserve nature with is culture; the only thing we have to preserve wildness with is domesticity' (Berry 1987: 143). Herein lies contradiction, contest and connection. The theoretical opposition between nature and culture is embodied in the works of those artists who have chosen to work with, and indeed in many examples *for*, nature. Together with the growing awareness and concern for the environment, since the 1960s certain conceptual artists have rejected the institution of the gallery in which to frame their works, instead choosing to make their mark directly within the landscape itself. Perhaps one of the most interesting examples of this within an urban landscape is the practice of graffiti – the anthropologist Schacter's discussion of which affirms

> the power of ornament to not simply reflect but to create order ... [disclosing] the distinctly ritual quality of the productive processes from which these artefacts emerge ... [inhabiting] the space between the ordinary and the extraordinary, that exists on the borderline of art and life
>
> (Schacter 2014: 10; 221–2)

Our relationship with landscape, and the effect we have upon it and leave behind us, is complicated: 'we see stability in its mute performance and flux in its unending variances' (Kastner 1998: 17). Berry writes of the human predicament being 'animated by profound connections and insurmountable divisions. The best Land and Environmental Art highlights this contradiction, probing the limits of artistic activity with the limitless tools of the artistic imagination' (Berry 1987, cited in Kastner, 1998: 17). Matilsky describes ecological artists as our contemporary shamans with their visions to restore the once vital balance between humans and nature (Matilsky 1992: 115). Lowndes discusses the early attempt by some conceptual artists to restore the relationship between artistic creativity and social reality in Europe and the United States in the 1960s (Lowndes 2014: 11–12). She cites Marioni, an artist and curator living on the west coast of the United States: 'it was an invisible decade; the work that was produced had low commercial value; relics, documents and photos of events, earthworks and installations – all works not made as ends in themselves. It was a vital era and the art world could hardly wait for it to pass' (Marioni 1988: 36). Yet it has not disappeared and, as Matilsky writes of ecological art, 'it is slowly beginning to make an impact as it provides solutions to environmental problems' (Matilsky 1992: 115). In her introduction to *Fragile Ecologies*, which discusses and portrays the interpretations and solutions provided by ecological artists, Matilsky traces the ways in which during human time artists have defined their environment; how, from the early celebrations of nature's cycle of growth, decay and renewal, 'art and ritual reflected the symbiotic relationship between people and the land amongst hunters and gatherers, and agriculturists' (Matilsky 1992: 35). This connection was lost, she argues, at the time of the industrial revolution and 'the flourishing of landscape painting represents a direct response to this schism. Like their preindustrial forebears, landscape painters communicated the spiritual and physical energies of the earth' (Matilsky 1992: 35).

This communication is still being practised by the landscape and contemporary artists of today, their works often being both 'agentive and transformative' (Pinney 2002: 135).

Inspiration, emotion, time, memory and walking

We have already referred to anthropological studies concerning walking and considered why people walk in Chapter 8. So it is interesting to note the manner in which walking or journeying has become a strong focus for many artists and curators, both for individuals – see for example, Anna Dillon (n.d.) and Alison Lloyd (n.d.) – and for those who choose to walk or journey with others as a means by which to explore 'processes of dialogue and negotiation' (Pope 2012: 57). An example of this is the Walking Artists network, which seeks to connect people who regard themselves as walking artists, those who regard walking as a mode of art practice as well as those in related fields such as cultural geography, history, and anthropology. Deveron Arts, based in Huntley, northeast Scotland, possesses no gallery or arts centre. Instead, its Director, Zeiske, writes: 'the town is the venue, working with the spaces the place has to offer' (Zeiske 2014: 91). The group works collaboratively with artists from around the world and the local community, bringing together 'artistic and social relationships in a global network that extends throughout and beyond the geographic boundaries of Huntley' (Zeiske 2014: 91). *Calendar Variations* is a series of practices by individuals working from a score by the artist Allan Kaprow in order to 'move from possibility to action, from score to performance (as) by understanding that Kaprow sought to develop social experience by artistic means, we become free to work with our centre and also (be) challenged to create common ground between participants' (Coessens and Douglas, 2011: 13).

Pope's work is regarded as central to the way in which walking as a method of art production has been rethought in recent years, Collier describes his group walks as 'deliberately subverting the Romantic notion of the solitary walker' (Collier 2013: 113). Pope's *The Memorial Walks* were created as a homage to W. G. Sebald, 'drawing on his use of walking and the stubborn insistence that the past would not fade from memory' (Pope 2012: 57). Invited guest walkers, among them those who wrote about landscape, the environment and memory, were requested to spend time with a painting of a local landscape from the Norwich School, the first provincial art movement in Britain, in which a tree was the main focus, and then accompany Pope on a walk around the East Anglia fenlands and farmlands: 'each writer would perform a recollection from memory of a tree. In doing so I had hoped that they might repopulate the countryside with images, summoned-up and made to live through the sheer force of a spoken-word description, as an act of defiance against forgetting' (Pope 2012: 57). Dillon, one of the guests accompanying Pope, wrote later:

> accompanying the artist into the countryside around Norwich, standing in the rain before a nondescript stretch of land, one realised that the chosen painting had become nothing more than pure atmosphere: it had been sublimed into the air and become the mere ghost of a painting, its afterglow … Art, we might venture, is just that realm where the atmosphere becomes visible … which infolds itself into the dark, which hints at affinities and correspondences across time, which evokes rather than narrates … may well be the art that most closely answers our sense of wonder or curiosity.
>
> (Dillon 2007: 23–4)

When we look at the emotions felt and what may then inspire an artist we find a number of aspects that both contemporary realist and environmental artists hold in common, including their perception or feelings about time and place. One artist exploring the boundaries of representational painting and time is Californian artist Gregg Chadwick. In an interview with Jeffrey Carlson, editor of *Fine Art Today*, Chadwick states his desire to 'break down the illusions of linear time passing and expose the co-existence of past, present and future' (Carlson n.d.). Art critic Clothier describes Chadwick as a literate painter; his readings and experiences within the landscape walked 'are processed through the work of the arm and the wrist, the hand and the heart' (Clothier 2013: 4). As the work comes into being 'each choice, each image, each gesture is informed with meanings, all of them so deeply interwoven as to be indistinguishable as single threads' (Clothier 2013: 4). In a discussion with Chadwick the importance of walking, inspiration and emotion become clear. He states that they 'are all deeply entwined in my artistic practice. Emotion is often an entry point into my art. And a balance of emotions is crucial both in my process and my paintings' (personal communication October 2015). For example when walking in Verona his sense of awe at the layers of history within the city and its surrounding countryside is 'tempered with a sadness of historical memory. As I painted these artworks of Verona I felt both the joy of human achievement evident in my subject matter and the anger and loss over the countless wars from the Romans through the Nazis and into our current era' (personal communication October 2015). He does not walk to escape:

> and not really to get to a place but instead to be in each spot as I take my steps ... Snippets of overheard conversations, the smell of lilacs, the crunch of my shoes on gravel, a blaze of light scattering across a shop window, passing traffic – all find their way into my storage banks of inspiration and reappear as colors of time in my paintings
> (personal communication October 2015)

Land artist Andy Goldsworthy writes:

> I cannot disconnect materials as I used to. My strongest work now is so rooted in place that it cannot be separated from where it is made – the work is the place. Atmosphere and feeling now direct me more than the picking up of a leaf, stick or stone ... a long resting stone is not an object in the landscape but a deeply ingrained witness to time and a focus of energy for its surroundings.
> (Goldsworthy 1998: 6)

He then goes on to describe how he repeats his visits to some stones and places, as they change according to season. He is interested in the 'binding of time in materials and places' (Goldsworthy 1998: 6). Stone and time are important for the walking artist Long too: 'I like simple, practical, emotional, quiet, vigorous art … I like to use the symmetry of patterns between time, places and time, between distance and time, between stones and distance, between time and stones' (Long 1980). In an interview fourteen years later he says:

> It's literally the same stones and the same surfaces of the world that people have always walked over and used. All the place names are like layers of history and different cultures. My work is just another layer on the surface of a world that has been shared by all these different generations, so it's really about continuity.
>
> (Long 1994)

Long's walking artist friend, Fulton, states that he walks on the land 'to be woven into nature … walking into the distance beyond imagination', his artworks acknowledging 'the element of time, the time of my life … (where) walking is the constant, the art medium is the variable' (Fulton 1995: 8–10). Describing walks as the kilometre stones of his life, Fulton writes 'each walk marks the flow of time between birth and death' (Fulton 2015). In one of the most important exhibitions held as yet of walking and artists, the 2013 *Walk On* exhibition brought together works from the late 1960s to newly emerging walking artists. Co-curator Morrison-Bell states the intention being 'for their paths to cross, so to speak, and for the viewer to experience, look or feel how an artist's walk could also possibly become the viewer's own, leading him or her to hitherto unknown places' (Morrison-Bell 2013: 2). In his discussion of some of the artworks, artist and co-curator Mike Collier writes that he feels many of the artists share an embodied or phenomenological approach to the making of their work:

> either – the way that they 'represent' movement through space (by walking), activating senses we sometimes take for granted … the way that they engage with an embodied experience of space and depth (what Merleau-Ponty called the 'flesh of the world') … the way that their work engages with others … making art is a practical application of phenomenology.
>
> (Collier 2013:73)

Artists and anthropologists sharing the same space/place

An exciting aspect of our anthropological study of the Pebblebed heathland is the way in which artists working in a variety of practices share this same place as us in what is frequently the sensual production of their work, and, as the reader will find, an often self-reflexive approach. This and other aspects lead us to note the now major discussion between artists and anthropologists as to the differences and possible similarities between the artist's and the anthropologist's gaze and means of representation. These concern fieldwork and the affective presence of the artist or anthropologist, and the importance of immersion, decontextualization, embodied engagement and the practice of repetition in order to 'know'.

Three important books edited by Schneider and Wright contain essays providing us with perspectives from both artists and anthropologists. For example, a detailed exploration of appropriated methodologies and subjects between the disciplines of art and anthropology, and the possible development of new practices, may be found in their *Contemporary Art and Anthropology* (2006). In *Between Art and Anthropology* (2010) Ossman's discussion of her fieldwork practice and the contribution of painting to developing anthropological knowledge (2010: 127–34) is but one of the presentations and discussions about work being produced within what Schneider and Wright describe as the inter-space between the fields of art and anthropology; they discuss the fragile nature of this ongoing dialogue. In *Anthropology and Art Practice* (2013) Schneider and Wright introduce the work of practitioners 'subjectively chosen' for their representation of 'particularly challenging and productive engagements with the shifting area between contemporary art and anthropology' (2013: 2). An excellent example of an artist whose methodology is akin to that of an anthropologist is that of Lang. Art critic Metken describes his exhibition, *Nunga und Goonya*, held in Munich in 1991, as being at first glance similar to a cabinet of curiosities:

> There are implements lying next to weapons, minerals and articles of clothing. Rock and colour samples can be seen, limbs of exotic animals, bark containers, a coal wagon, grass and feather capes. Good, one says to oneself: a somewhat sporadic anthropological collection, extremely widely deployed and embedded in its natural and social environment … however this is no systematic collection.
> (Metken 1991: 34)

What is on display is a confrontation between cultures, between nature and culture. Helmut Friedel *et al.* (1991) write that in his various works Lang attempts to collate pictures of human encroachment on nature: the methods used in so doing always remain the same. Findings that evidence human activities and interference are collected, ordered, described and examined as to their meaning. Researching the historical and local contexts, the 'biographies which come to light from the finds play just as an important a role as the actual remains and traces he has found in the completed picture (he) creates' (Lang 1991: 6). In *Nunga und Goonya* Lang is mourning for the lost cultures of Australia's indigenous population, the Aborigines, and it is the latter's view of the white settlers who regarded the Australian continent as *terra nullis* that he presents (Lang 1991: 8). Yet as Metken remarks, Lang 'remains the white artist who takes his findings back home with him and uses them for his purposes … which does not exclude any amount of commitment, not to mention dismay over the injustices continuing till the present day' (Metken 1991: 36).

Among the heathland artists discussed below, Trenchard's approach and works, both on-site and in her Master's portfolio, are a good example of a non-textual, visual production that is often anthropological in its essence. This is important in view of the dominant textually documented and descriptive negotiation and representation of place, emotion and aesthetics. Schneider and Wright correctly emphasize the difference between 'participation' and 'collaboration', stating that 'they are charged forms of rhetoric that have been subject to much critical scrutiny' (2013: 11). Participation involves a 'whole constellation of different degrees and conceptions of agency and control at work … and the specific complexities of particular contexts for collaboration require acknowledgement' (Schneider and Wright 2013: 11). Perhaps it is also important to state that, although anthropology has been critiqued for its anti-aestheticism, Geismar and Empson (2004) argue that anthropologists such as Edwards (2002), Gell (1998) and Pinney and Thomas (2001) have acknowledged the power of the visual both 'in its own terms, as well as through more academic discourse (in which) the importance of the visual (is) a crucially material category in vital interaction with socio-political, economic and cultural contexts' (Geismar and Empson 2004: 44). There is a difference too between anti-aestheticism and 'beyond aestheticism'. As Pinney states, 'it is not the efflorescence of words around an object that gives it meaning but a bodily praxis, a poetry of the body, that helps give images what they want' (Pinney 2001: 161).

Heathland arts

A painting of the heathland that few members of the general public have seen hangs in the Officers' Mess at the RM Commando Training Centre, Lympstone. Painted by Margaret Dean in 1981, it was gifted to the Royal Marines by the Oxford Architects Partnership. It is a large oil painting, about two metres wide.

Margaret says she had a choice between the Exe Estuary or Woodbury Common and she chose to paint the latter: 'I chose Woodbury because I liked the tangled intertwining of the brambles and gorse bushes in the foreground, which helped to push back the furthermost part of the Common to give a big spatial sense to the place' (Margaret Dean, artist). Arguing artistic licence, Margaret moved some of the trees in the distance as she felt she needed them to be differently distributed for the picture's sake. This has been picked up on by some observant commandos who know this area very well indeed as it is the landscape they train in. Amusingly she says: 'The commandos also say that they can see figures moving around in the undergrowth – I can't understand how they can see things that are not there – and I should know!' The initial work, mostly drawings and notes, was done from inside her car very early in the mornings: 'I was very nervous during

Figure 9.1 Margaret Dean's painting of the heathlands hanging in the RM Officers' Mess at Lympstone Commando Training Centre

this time because it is so isolated and I used the car as a safe place to be. There are quite a lot of people seemingly alone and I didn't want company at that time'. The painting was done in her studio and took her about six months to complete. Margaret says she loves the heathland area for its light, undulating terrain and the sense of being close to the sea and the beach pebbles.

Margaret Dean's painting is unusual in another sense. We noticed on visits to exhibitions by local artists a striking absence of landscape paintings of the heathland. Favoured scenes were of the area surrounding it: views along the River Otter or of the Exe estuary, or of the sea and the coast. This is no doubt because the heathland itself is not sufficiently picturesque, definitely not the kind of bucolic English scene we see depicted in the work of Gainsborough, Constable and others. This is not to say that those practising the creative arts are uninterested in the heathland. Indeed many visit it and derive inspiration for work and practices of another kind. During the course of our fieldwork we talked to a local poet, an actor and theatre director and performance artist, an acrylic artist influenced by Jackson Pollock and a dancer. Their approaches, methodologies and artworks are as individual as their relationships with the heathland. We summarize a few points here.

For Barbara Farley, the poet, the heathland is a place where she walks regularly and takes picnics. Each walk she describes as being different in terms of the plants and wildlife she encounters: butterflies, dragonflies, etc. The heath invokes pictures in her mind and to her the heathland has different personalities in relation to seasons and places within it. Although she knows the heathland is managed it feels wild and untamed to her and is a spiritual place. She describes herself as having a photographic imagination. After walking on the heath she carries away pictures in her mind and these in turn enter into her poems, in which she paints a picture using language:

Watching For Nightjars
Barbara Farley

we gathered in a clearing at the edge
of the heathland on the cusp
between day and night

a sudden thunderstorm in the afternoon
had made the air clammy
now wrapped us in a cobweb shawl of moisture

there was a sense of anticipation
a few people spoke in hushed voices
but mostly there was silence

our guides led us down the track
towards the heart of the plantation
where we split into two groups – I took the lower path

we left the pebbled track

made our way downwards towards a clump of scrubby pines
the way became sandy beneath our feet

we walked with concentration as if
carrying a bowl of some precious liquid which we were
afraid to spill

we stopped beside a tree which stood alone from all the rest
somebody coughed
a violation which tore a hole in the flimsy fabric

of the dusk   the seconds   stretched
until I was sure time would break as we floated in that dark
lake of our own isolation

then it came – a churring like a thumbnail
drawn down a metal comb – rising and falling
somewhere out of sight

and then an answer   our island filled with noises
we turned   it was impossible to tell
from where they came

about our heads three long-winged shadows
wrote their cryptic messages
against an ever-darker sky

at last the sounds ceased altogether   ever
in silence we returned to where we'd started from
I saw a footprint in the sand

and knew that it was mine
all the way home   and for a long time afterwards
I felt

as if I cradled close to me
some tiny fragile form whose warm heart
is beating still

Jon, the performance artist, who had come to the heathland for the first time when we talked to him, was interested in 'mining the landscape for stories and information and feelings about the people in it'. He hopes that by linking the ecology and history of a landscape and presenting this in a performance people will value it more. He describes the process of entering into an unknown landscape and trying to understand it as being 'heritage art'. By walking he attempts to understand the atmosphere of the place, 'the things that people know about without even knowing that they know about them' (Jon, performance artist). Place makes things happen in the mind and he wants to re-mythologize space. Observing a gap in the ridge to the east of the heathland he suggests a potential story, an audio walk: 'I might decide that that ridge, that hole in the ridge line was caused by a giant who awoke from beneath the ground and took a lump of it for his tea' (Jon, performance artist). He is fascinated by the Iron Age hill fort of Woodbury Castle, viewing it as an ideal performance stage, describing how a number of little promenade stations could be allocated so that people could come in and experience different types of performance: 'you could do things in the trees; you could make little hollows in the bracken; you could walk people around the site and give them a sense of that fairy feeling because these places do have their own atmosphere'.

Caroline, the acrylic artist who walks regularly on the heathland with her dogs, feels that the ubiquitous pebbles have subtly influenced her work because she's interested in their shape and colours. She goes on to describe the energy of colour:

> I'm just such a colour person. I love the energy of colours and what they do. They're next to each other and so two colours will talk to each other but they'll talk differently. Even if you're not working representationally at all, those things [the pebbles] are underneath somehow.
>
> (Caroline, artist)

The experience of walking on the heathland influences her work indirectly in another way: 'it's an important part of the reflective process, a reflective time for me'. For her the heathland conjoins emotion, imagination, movement and memory. It embodies a different time and place in which, restored, she has been able to return to her life, to her artwork.

Michelle, the dancer, for several years has been exploring dance and yoga in relation to nature. She usually comes to the heathland at weekends or, in the summer, in the early evening and tends to use one of three sites to conduct her personal movement exploration. These are the Woodbury Castle area, an area out on the heathland in the woods not too far from the castle, and the opposite side of the road that looks back on the River Exe. She describes herself as having an instinctive feel about where she wants to be on any particular occasion – either which side of the castle or in the castle:

> If I'm in the castle I'll probably be drawn to work with a group of trees, or a particular tree, or between the trees. The weather conditions will affect this decision because this is going up every week during the year and if it's very, very cold, I probably wouldn't do so much static exploration; I would do something a bit more physical.
> (Michelle, dancer)

So first she arrives in an area she is attracted to. She may move around quickly but often she is still and allows her energy vibrations to connect in with the actual place: 'So I'm much more open to the sounds, to the atmosphere and the conditions of the place rather than coming in and saying "I want to dance on or in this place". It's allowing the space and the place to invoke in me some kind of movement response'. Believing that thought, feelings and movement are all interconnected, Michelle responds to the chosen place and describes this response as a more open way of being in the environment than if she was walking, where she feels she is more of a spectator: 'Movement will come through. By spending periods of time in a smaller part of an environment, I tend to have perhaps a deeper relationship. It's a slightly different experience to just walking on by where you're constantly stepping from one place to another place'. In this way, the actual place itself, rather than the landscape generally, is of most importance, as she feels the latter describes a flat picture or terrain. In 'place', she allows herself to become part of it and the feeling of the ground beneath her, its textures and smells.

Michelle has taken part in a piece performed during Heath Week 2010 by Landance, an organization that runs workshops in contemporary dance, music, visual art and film that lead to performances in the landscape of the south-west (http://www.landance.org.uk) and has permission to conduct workshops in the castle area (this is just one of the many sites that she uses locally). She states that the premise that she works outside certainly has the same ethos at each site but that each also brings its own qualities. Also, each person who comes to the workshop has their own personal response, their own personal relationship to the space. Sometimes people have driven quite a way to come to the workshop and on such occasions Michelle conducts a warm-up that may entail feeding in suggestions about moving different parts of the body: 'I sometimes do a little movement, a ritual, where we all just stand together and do similar kinds of movements but not having to do it exactly how I do it. This is a way all of us can be together as a group and also of just coming in to the body'. The participants are moving outside in a public place, moving and experiencing the space and different weather conditions in a way that is likely to be dissimilar from how other people, such as dog walkers, are experiencing it. Michelle says that people seem to like having this structure first and that it is a helpful transition to developing an individual response where people move off, not far from each other so they can be themselves but still have the sense that they are part of a whole group that is moving. Each workshop tends to be about three hours in length and coming together in this way at the start is supportive. During the course of the workshop, often at its end, there is a space for artistic or personal response: 'People might write or draw or create a little bit of environmental artwork'. Some of these responses may be viewed on the 'Moving Naturally' website (http://www.movingnaturally.co.uk).

Thus we can see how moving the way she does in the heathland locations is part of what Michelle has described as her journey of human embodiment. It involves the physical movement of yoga, dance and other movements that are free from being stylized, the use of senses more than just the visual, together with an opening of the emotions and thoughts. These are not processes that are undergone separately or felt individually but a sensitizing embodiment in an empathetic relation to the environment moved in.

In the following section we consider in much more detail the work of two local artists whose work has been directly inspired by the pebbles and the heathland landscape itself. During an unguarded moment while talking to the assistant manager of Black Hill quarry, Chris remarked that the pebbles that were to be annihilated by crushing and turned into aggregate were rather beautiful and aesthetically pleasing. Had he kept any that were particularly interesting? The assistant manager's jaw dropped with an expression of total disbelief. He eventually commented 'Well, you can go back to the Stone Age and live in a mud hut but we need these materials to maintain our prosperity and way of life'. That, if you like, is a functionalist view of the qualities of pebbles. We now examine them as material media of both artistic agency and aesthetic appreciation.

The uniqueness of pebbles: the story of the beach artist

Having retired early due to ill health Barbara Hearn started to paint when she was recovering. One evening when walking at Ferrings, a beach on the English south coast, she looked down, saw the pebbles and thought 'Gosh, that would make a better painting than the one I'd been trying to do'. She took a photograph but found she could not paint pebbles very easily from a photograph and so returned and collected some samples. Since then she has painted pebbles from a number of beaches and now, living in Budleigh Salterton, paints the pebbles from this location. Her works, then, are based on the border between land and sea. Barbara has sorted and arranged pebbles since her childhood: 'A very early memory of mine is having shops on the beach and having rows of pebbles and shells and selling them to people. That was my shop. Everything arranged on that beach very beautifully'. She finds that each beach visited is different and that although they have more or less the same kinds of stones, the proportion of colours to be found varies greatly. Each pebble is regarded as being special: 'I haven't come across one that's not special but there are no other pebbles like Budleigh Salterton ones, are there? That's the only place where I've found this particular shape. They're not round, they're flattened. They are a very specific shape and form'. She explains how there are many different ways to look at a pebble, as there are several different angles to view it from and one side of the pebble may be different to the other. Also, when the pebble is turned round it can look very different and this can result in a particular pebble appearing in several paintings without ever looking the same. This handling of the pebble is pleasurable for her: 'I like the feel of them, the tactile sense'. Some of these pebbles are very smooth; others are more textured: 'The Budleigh Salterton pebbles are interesting because of their speckled nature and the patterns you get on them'. Barbara herself has favourite pebbles. She particularly likes pinky-coloured ones that have many speckles on them but has found, when conducting workshops and when serving as an artist in residence in a school, that the sort of pebble she thinks most pleasing will not interest someone else. For her, the combination of the pebbles' elliptical and repetitive forms, together with the variety of colours and patterns, is fascinating.

Colour is of great importance to Barbara and this is why she describes the Budleigh Salterton pebbles as 'a joy to paint'. Some of the first pebbles she painted were bright orange, but she then realized that these were quite unusual ones. Now she focuses on getting an initial correct representative balance of colours of the beach's pebbles in her paintings, although she says that she does tend to put in more white and black pebbles than you would find proportionally on the beach because it helps 'make' the picture. She believes there are more greys, pinks and purples than orange, black, brown or white pebbles on the beach at Budleigh Salterton and feels that the Pebblebed Project's attempt to colour-code the pebbles in the archaeological excavations must be a difficult task:

> Because you get that orange with the pink, don't you, and you get a deep sort of brown with purple in it and then you might turn the pebble over and find there is a corner that's grey with bits of orange in it. Even the red ones have got that brown in them. So they are very unusual.
>
> (Barbara Hearn)

It is interesting to learn of how Barbara works colour when she is painting. For example, for the pinky-coloured pebbles she uses a permanent rose that is deep pink in colour and adds burnt sienna, which is warm mid-brown: 'Now those are two colours you would never have on your palette together but when you are painting Budleigh Salterton pebbles this is necessary'. She remarks how the soil on the cliff face itself is red sandstone, a red-orangey colour, and this is maybe why she adds burnt sienna too in order to get the correct tone of pink.

Prior to painting Barbara does not have an arrangement in her mind but chooses a number of pebbles of varying size and colour. The pebbles are then wetted as this exaggerates their colour. Chosen pebbles are laid out on a tray: 'This means the spaces between the pebbles are interesting … on the beach they would be overlapping so what I do is quite artificial; it's almost like an abstract'. Next she chooses a pebble that she really likes and this becomes her focal point. She draws its shape but does not paint it at this stage. Then she takes another pebble whose shape she finds pleasing and draws that. Pebbles of the same colour are rarely placed together. In this way a small composition grows in the picture's centre, with an intention that there are echoes through the whole painting, with pebbles placed strategically so the eye travels: 'What is so very interesting, which I found from teaching children and adults to paint pebbles, is to find the composition pleasing is actually much more difficult than you would think and I normally now encourage people to find three pebbles that go together to start their painting'. Once this central composition is flowing she then starts to paint, having decided what her palette is for the beach; she normally has no more than four basic colours that she mixes. She tries to create what she describes as a limited palette and states that even bright orange pebbles contain four basic colours. A note of the colours used is made: 'I will write down the four colours because I may not finish the painting at that sitting, so I need to remember'. Textures are created when necessary and this may mean adding salt to create a clump effect or adding salt to a pearly mix when it is very, very wet. To get the subtle mottled or speckled effect, table salt is put in to the wet mix as it fixes the paint, and then a toothbrush used to splatter the speckles. Candle wax is used to create the white lines on black pebbles. In order to show the direction of the light each pebble has a light side and a dark side:

> While it is still wet I'll take a brush and put extra water on where I want the light bit using gravity to hold it, to let the paint drag down. And then if I'm not happy with it as it is drying I will add more colour. I fiddle around. It's fun to do, great fun. When people first do it they squeal!
>
> (Barbara Hearn)

Once she is satisfied with the centre, she slowly builds up other pebbles: 'What I will try and do is to get the eye to flow round the painting and that's where the Budleigh Salterton ones are useful because you can turn them on their side and they will be long and thin and they will make arrows for the eye'. Next the spaces are filled in with little pebbles whose colours will either grab attention – 'look at me' – or blend in. In this way she makes judgements regarding placement, colour and flow. The final piece of work is to add black spaces: 'Sometimes I just scribble with a very soft pencil and then wash it in with water and the graphite will fix. Sometimes I use a charcoal pencil; sometimes I'll use the colours already on the palette, all the oddments to fill in the spaces over the pencil to blend it nicely' (Figs. 9.2 and 9.3).

All of Barbara's pebble paintings are created in watercolour: 'it has to be watercolour because as the watercolour is drying, I'm dropping colour in and fiddling with it'. She runs a pebble-painting course named 'Painting texture': 'It teaches them the techniques of getting different textures and how you can achieve that with watercolour. You couldn't do that with oils'. She also creates a dark background as this has the effect of throwing the pebbles forward: 'If you keep it fairly gentle behind the pebbles the picture does not look so dramatic'.

When Barbara initially tried to paint pebbles it was from a photograph, but she found that this did not work very well. The essence of the pebble was not captured. She needs to hold the pebble while she is painting it: 'Yes, if I don't handle them, I don't paint them anywhere near as well. It's interesting. There is something about the feel of what I've got in my hand'. Barbara has painted Budleigh Salterton pebbles so often now that she finds she has a quick response to them: 'I think it's capturing some kind of emotion from them'. In turn, she describes herself as not being conscious and being lost to the world when she is painting: 'I wouldn't eat, I'd forget to drink; it's like a bit of meditation with that bit of paper and the colours ... very content with myself and not really worried about what was coming through, not thinking about what I was painting, being at one with what was inside me'. Importantly, Barbara only paints when she is happy: 'I can't paint when I'm sad, when I'm not feeling good about myself I cry when I try to paint'. Her paintings then, may be considered to be happy paintings and she believes that comes through: 'Some people said that when they see my work; not everybody but with some people it resonates. I do know when I'm painting well there is a joy there'. However, she does not want people to make conceptual associations or read meanings in to her pebble paintings: 'I want people to make them their own really'.

It may be seen that Barbara's relationship with each pebble painted is personal. Each pebble is held, its essence captured in the painting, the colours used echoing the pigmentation found in the pebbles and their environs, and the colours in her mind's eye becoming a corresponding repetition of what is seen and felt, a repetition of form but uniqueness of being. She becomes attached not only to the pebbles she paints but to some of her paintings too, and nearly cried when she sold her first as she could barely part with it. Her absorption and contentment when working are part of this process of the capture of colour, essence and a making of something that she hopes will bring happiness to others too.

The Pebblebed Project artist's story

Our next artist, Priscilla Trenchard, is another who has had a long-term fascination with pebbles. She too is interested in sorting and arranging the pebbles and takes pleasure in the tactile experience they give her. The way she works is very different from Barbara's methodology, however, as are her completed pieces of work. First we will look at Priscilla's

Figure 9.2 Pebble painting 1. Painting by Barbara Hearn

Figure 9.3 Pebble painting 2. Painting by Barbara Hearn

experience and feelings about the Pebblebed heathlands. Then we will discuss the work she produced for the Pebblebeds Project.

As a child Priscilla lived in the coastal town of Seaton and brought pebbles home, and since then – 'Always, wherever I go, even in the US, I've collected pebbles'. She has 'gone past' the heathland much of her life and visited Woodbury Common as a child but it is only recently that she became so involved with the Common, and this was through her university course work. She and her partner, a cultural ecologist, wanted to work on something together that was related to people's reactions to things in the landscape. At first she was going to make a cairn on Budleigh Salterton beach and see whether people added to it but then a friend told her about the dig on the heathland: 'I had no idea who Chris Tilley [the project director] was but I wrote to him and I met him and he said, "Join in things", and it was great; my Master's thesis was written up about the dig'. Priscilla illustrates this engagement in her contribution to the book *Between: Ineffable Intervals* (PLaCE 2012: 64–70), a collaborative work by thirteen landscape artists.

Like several of our interviewees Priscilla remarks that there is little in the British landscape that has not been 'touched' and so although she finds the Pebblebed heathland to be a natural place, it is not wild. In fact she describes it as being 'uncomfortable' and 'painful' when walking through the gorse: 'I was like a little pin cushion. I think it's slightly bleak, quite a moody landscape; I don't think it's altogether inviting; it's slightly aloof and I found it quite intimidating initially'. This initial experience came after she attended the archaeological excavation and was invited to work in the landscape; her feelings of intimidation were perhaps not just to do with the heathland itself but also because she was joining a group of people whom she did not know: 'I didn't know Chris, I didn't know anybody; who were the professionals and who were the volunteers. It was more than just landscape, it was personalities as well'. She remarks that this is her first experience of being on an archaeological excavation and has never been anywhere where pebbles have been scrutinized in so much detail.

Soon she started to appreciate the heathland, finding the actual process of getting to the excavation sites themselves as an enjoyable experience: 'The process of going there, the process of going with the whole group, the same trails every time and setting up the site; I really got to love it'. Priscilla finds 'repetition' to be of importance, returning over and over again to the same place in order to get to know it: 'That's what I like about Colaton Raleigh Common. I look forward to seeing the next ridge and then you get to where Tor Barrow is (a Bronze Age pebble cairn) and there is the whole sense of arriving that I like'. She finds that when she drives across Woodbury Common she now sees it differently; that it is not necessarily a love with the place but she is now intrigued with it: 'It is the familiarity with it that makes me want to go back'. The heathland's bleakness, gorse, heather and trails are now appreciated.

In her writing on the Pebblebed Project website Priscilla describes Tor Barrow as like being in Avalon: 'I felt we were in a special place. It's unfriendly terrain and yet it felt very powerful. Down below was all the greenery and I did feel we were up somewhere wonderful, and the connection with the group as well; it was very, very interesting'. Her feelings towards another excavation site at Aylesbeare are not as warm: 'I was quite grumpy there. We all sat like we were on a beach, looking at the beautiful view but, I don't know, it just didn't have the serenity that the other places have'. She explains that this location was close to the road and she could see a few buildings, making it more connected to contemporary life than the Tor Barrow, which felt more special with a sense of belonging to the project team. When contrasting her experience of being on the Pebblebeds with Dartmoor, she describes them as not drawing themselves to her the way Dartmoor does. When asked if she found the heathland changed in character when it was sunny she describes how she herself changes in character when it is hot: 'I'm not very nice, so, I was really happy with the weather up there. There's always a breeze. I can't bear still air. I need to feel movement and air so I was very glad that it was dull up there'. Priscilla likes to know there is what she calls an 'edge' to the landscape and describes how she felt claustrophobic when living in North Carolina with its mountains: 'The one thing about being in Britain, the edge is never that far away. We were high up and down there was all this lusciousness and the lovely aspect of High Peak and that whole ridge that goes along the top'.

Besides her excavation work, the only recent occasion she has walked the Pebblebed heathland was with Chris, soon after a big wild fire had swept through. She found the blackened and charred landscape interesting and describes it as being like a drawing:

> It was like a living drawing to me because of the black lines where the bracken was just starting to grow. The new ferns looked extraordinarily like black spirals but when I went to pick them, they were juicy inside. Despite being charcoal burnt on the outside, they were still alive inside. It was very odd to be somewhere that was completely black. After a while it was a bit depressing because you knew why it was black as well. The sad thing was we found cooked

eggs and all the things that would have been alive on the surface – if they couldn't fly, they were cooked.

(Priscilla Trenchard)

Priscilla remarks on how the fire had changed the structure of parts of the landscape: trails had vanished and this made map reading difficult: 'And, of course, it smelt still of fire and I actually love that smell, the smoky smell but again, you know the damage that has been caused'.

The pebbles themselves Priscilla finds interesting, and feels that they are reminiscent of her childhood where she collected them in Seaton:

They are tactile, very basic earthy connections with the land I think. The patterns in them, the grey ones with just white stripes. It's fascinating to me. It's got a graphic element so sometimes I just collect bundles of those; ones with holes in. It's like they all have a little mystery to them.

(Priscilla Trenchard)

She remarks on how people look for pebbles when they are on the beach at Budleigh Salterton, and she does the same: 'I'm not looking up at the view but down at the pebbles. Some of the ones that aren't exciting can still be of interest to me'. The shape of the pebbles and their smoothness is, she believes, what makes them special, and although she likes their colours she describes the scope of those colours as being limited. Their magic lies in the holding of them. She says that although some people read things in to the pebbles, she prefers not to: '"I can see a witch's face; I can see a tree"; I tend not to want to see images in this way. I tend to look at the pebble and enjoy it. This colour is next to that, and the pattern and the way when you turn the pebble it changes. It may be mirrored on the other side, it may not'. As a child Priscilla always had a matchbox of objects in her pocket: 'It held either a stone or a bug or a piece of something, a connection. It was like some other land, some other place I could go but I think it takes me back on a journey through history, it is something which is solid and re-assuring'.

When we look now at the work Priscilla did for the Pebblebed Project dig and her Masters degree we will discover some of the ways in which her heathland experience and the feelings associated with it have helped influence what has been produced and how.

Having taken part in excavation work at Tor Cairn Priscilla felt she had to break away and commence her artwork, but she felt quite guilty at leaving the group to do this. It made her question what 'work' is, whether it means producing something or if having an 'idea' is also 'work'. She had never worked in a landscape before and knew she did not want to bring paper in to it. Another woman on the dig knew quite a lot about weaving and this interested Priscilla: 'I thought, okay, I'm only here so many days; I can't learn to weave properly in this time; I can't actually do a proper weaving'. She decided, however, to weave a container for the pebbles (which would have been carried by the prehistoric people to the cairn in this manner), and initially tried working with willow. This she found to be a frustrating process, mainly due to her lack of knowledge about which season such materials should be gathered and her attempt to create a structure that was regular in shape. The next decision was to use the materials that were more to hand – the gorse and heather. She walked down in to the valley to harvest materials that were manageable and pliant and then considered that this too was not quite right: 'I thought, no, I'm going to deal with what's here and what I can do with what's here'. And so, using the rigid gorse and heathers, she commenced by putting a few things together, tying them initially and then pushing in more materials. She describes the weaving as taking on a life of its own: 'I wanted to make a basket and it just kept growing out'. Eventually she placed it on the cairn: 'I put it in the landscape and I thought, "Oh, it's done". It looked like a woven flame so I called it "Woven Flame"'. To her delight the other people working that day came to look at it, each one of them leaving a stone inside of it. It was as if a form of ritual had been established. When asked whether she felt that this was akin to a giving back to the landscape, Priscilla said it was: 'Because it was very much of the stuff that was there, of course it looked at home there'. The roots of gorse mimic the way the gorse grows above ground and this lent Priscilla the notion that the weaving was mimicking back at the landscape. To an observer, looking at this Woven Flame, one can feel that it not only mimics itself back to the landscape, it also looks back at the observer.

Priscilla says she felt this too and put her camera inside the basket to take a photograph looking out. She has gone back to the site to see how this weaving has changed with the seasons: 'It's not woven properly so it will eventually collapse I think. So I go like a pilgrimage to the site'.

Priscilla's fascination with the fire at the dig (the excavation team had found traces of fire and burning on the old ground surface on which the cairn was built) also led to her own work with fire and smoke (having invited the fire brigade to check her home beforehand for possible dangers): 'Fire, watching flames seems to be a very important thing ceremonially in lots of cultures, and in Christianity you have baptism by fire. It's

Figure 9.4 Woven Flame 1

Figure 9.4a Woven Flame 2

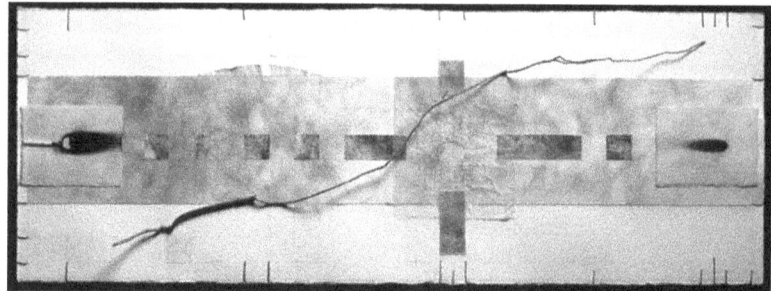

Figure 9.5 [Trans]figure monoprint. By Priscilla Trenchard

a very important element to live with so it's very significant'. One of the things she does is to try and capture the smoke; she tries to draw it and has found that many of these images appear to contain human form. She has a sink full of water into which the carbon, from candle smoke on paper, is released onto the surface of the water. The carbon then reassembles and cracks and this 'drawing' is lifted on to another sheet of paper where it is held in a more stable state. 'It was very interesting to capture material disappearing in to what seems like nothingness' (Figure 9.5) (Priscilla Trenchard).

The tonal colours of both the pebbles and the soil from which they come are of immense interest to Priscilla and she has used the archaeological grid from the Aylesbeare site to make colour classifications:

> The whole thing about the soil, to me, is just wonderful, the tones, and I collected all the different coloured earths that came up, even the burnt bits. Chris let me take samples and I ground them down and have them all out and I may make pigments out of them, at some point. I'd like to impregnate something with the pigments, you know, it's been fascinating; I can't tell what else will come from it.
> (Priscilla Trenchard)

When asked if she thinks colour can evoke emotions, she agrees and states that formerly she used a lot of colour and although people may respond more easily to colour because of the emotions it may induce, her work is now more monochromatic and subdued.

Of her pebble paintings she describes much of it as being to do with a calligraphic mark:

> Having done calligraphy and learnt to manipulate the drawing instrument, the way you hold it, there is something wonderful about the pebble shape, the way you make it. So I like that gestural

flow, that way of working; when you are drawing them as pebbles it's to do with depth and overlay and counterplay.

(Priscilla Trenchard)

She says that most drawing is about light, not line, and when trying to make a mould of a pebble this created problems:

> Because when we look at it in our heads we actually create an outline. A pebble doesn't actually have an outline like that. If I was drawing it from an actual pebble, it would have to be about light hitting a surface so I have abstracted it to a movement about line. It is very rarely that I would draw a line that meets up the other way. I do multiple lines.
>
> (Priscilla Trenchard)

Chris asked Priscilla to draw the pebbles in a section of the cairn using an iron grid. She did not enjoy this as, like the artist Barbara Hearn, she finds a pebble has no edge and there are several angles from which it may be looked at: 'Every time you moved your head to one side of the grid, you saw something different. It was doing my head in. In a more patient life twenty years ago I would have loved doing that, now it's a "no, I can't"'. As has been stated, Priscilla does not read things in to the pebbles and her work is not about drawing things as they are but from what she can extract from the thing – pebble, smoke, earth: 'It's more about expression'. The iron grid used in the technical process of drawing an archaeological section was used, or subverted, by Priscilla for another on-site artwork. She collected pebbles of different colours and placed them in the grid. When wetted they came alive, creating a dazzling display (Figure 9.6).

The methods used in the excavation work were found to be similar to those Priscilla often uses when making an artwork – working in layers and then allowing things to reveal themselves or not. For her some of the layers are in the repetitive journeying to a significant place, the archaeological site; the marking and making of a pebble colour grid for example. In one piece of work the marks made by indentations and other things in the soil are worked on and layered, the different images making up the composition: 'It's why you can't read it as being on Woodbury Common'. Other layers have been formed when she has left fabric with pebbles on it in the landscape from which a natural print forms from the debris and the mud (Figure 9.7). Another attempt at layering was when a fabric sheet with pebbles laying on it was successively splattered with

Figure 9.6 Pebble grid. By Priscilla Trenchard

Figure 9.7 Pebbles prints on cotton. By Priscilla Trenchard

muddy water. Instead of a material layered effect, it became ghostly in appearance, which again can be seen as a kind of layering, on this occasion the presence of past time. When she writes about her work now, making artistic statements, she describes the way she works as being an archaeological means of working with landscapes, materials and fragments: 'Putting things together so other people can make the connections and frequently the way I put paint down, it's surface on surface'. In her Master's exam she displayed the materials from which she had worked, placing the ground-down muds, charcoal burnings and sieved ash sections in jars – this in effect became the journal of her work.

Whereas Barbara finds background to be of great importance in her paintings of pebbles, Priscilla does not: 'For me it is a physical thing. It's not just the look of the pebbles. It's about it making a print and what will I do with that? I like working with the elements; it's all about the experimental, allowing things to happen, which I can't predict'.

Conclusions

We have seen in this chapter how the notions of repetition, colour, the tactile nature of pebbles, the pleasure in the visual aspects of a pebble and being in a landscape of pebbles, and the gestural movements used in painting are crucially important to our artists, an embodied and material relation that provides affordances for their greatly different works. Those working in the arts clearly have varying degrees of engagement with the heathland and its pebbles. A number of ideas have come forth – movement, time and memory, the embodiment of the human with the landscape, pleasure and discomfort, and a sense of belonging in the landscape. As well as feelings and emotions there is the question of consciousness and the unconscious processes that also may be at play here, together with their role in establishing such feelings and emotions. Essence, expression, holding and layering, memory, movement and walking have been seen to be held in varying degrees of importance, as are the artists' attachment to or engagement with the landscape and place, which have either subtly or strongly influenced their creative activity.

10
Fishing and the watery pursuit of 'pets'

In the UK there exists a fundamental division between coarse fishing and game fishing in freshwater recreational angling. Coarse fish are opposed to game fish in a rather neat binary opposition. The former are species such as carp, tench, rudd, bream, pike and perch, together with a host of others. The latter, game fish, are primarily just two species: trout and salmon. A social division mirrors this difference in the type of fish caught. Coarse fishing has traditionally been associated with the working classes and game fishing with the middle or upper classes, although such a division is now fragmenting. Types of rods used and techniques also differ fundamentally, as do types of bait. Coarse fish are taken with 'natural' bait that may be any mixture of bread, worms or any number of other materials (see below) whereas game fishing is undertaken by using artificial and preferably hand-made flies with a myriad of forms (for a discussion of the semiotics of fly fishing see Van Den Broek 1984; see Douglas 2003 for a discussion of the gendered and class basis of fly fishing). Coarse fishing in which the fish are kept in nets and then returned to the water, irrespective of size, rather than being eaten, is a peculiarly British cultural tradition and as far as we are aware does not occur elsewhere in Europe apart from Ireland.

Fishing on the heathlands is exclusively coarse fishing and takes place only on the Squabmoor reservoir, located in the southern part of the Pebblebed heathlands. This reservoir was constructed between 1864 and 1867 to provide drinking water to the rapidly developing seaside resort of Exmouth, replacing local boreholes providing brackish water (Delderfield 1948). This supply was replaced in 1909 by borehole

Figure 10.1 Squabmoor reservoir looking north

supplies from Dotton, near to the River Otter. Since then this facility has been used solely for recreational purposes, principally coarse fishing.

The reservoir, approximately triangular in shape, is 500 m long and 250 m in width at the widest point.

The dam at the southern end has drowned a shallow valley with a stream flowing south through it to reach the sea at Budleigh Salterton. Covering four acres, it ranges in depth between 2 m at the shallow and narrow end to nearly 9 m by the dam wall. There is a small overflow channel at the western end. Today the reservoir is managed by South West Lakes Trust. Daily or season fishing permits are available for purchase. Around twenty-five persons have annual season tickets; others purchase daily or twenty-four-hour permits on an irregular basis. Coarse fish present include carp, crucian carp, tench, bream, roach, rudd, perch and eel. There are also a few trout left from an initial stocking of the reservoir as a trout fishery. There are fifteen swims (fishing places) around the narrow end and both sides of the reservoir, together with others along the dam wall. So there is a maximum capacity for about fifteen persons, or more if they share a swim. Usually there are no more than ten or so present. Fishing is permitted throughout the year and by day or night. There are small car parking facilities off a minor public road crossing the heathlands 500 m to the north and next to the dam wall, approached by a rough track from Dotton Farm 750 m to the south-east. Regular anglers prefer this place because they can keep an eye on their vehicles, prone to thieves on the other car park, which is out of sight.

This tranquil and beautiful spot on the heathlands is popular with not only anglers but walkers and also cyclists and horse riders. There used to be a path going round the entire lake but it has been blocked off on part of the western side after an accident when a disabled person slipped off the muddy bank into a considerable depth of water and had to be rescued. Groups other than anglers are not permitted to use the track running beside the reservoir on the eastern side because of erosion but follow a higher track running along the valley side with views across the water. The RM do not train in the vicinity of the reservoir, although their presence on East Budleigh Common to the north can often be heard. The main problem the anglers had in the past was from mountain bikers speeding down the eastern side of the reservoir, a route that has now been blocked off with stiles. Their contemporary problem is with dog walkers exercising animals off the lead and actively encouraging or permitting their dogs to go into the lake. This is the main potential source of conflict today.

The principal user group consists of local anglers from Exmouth, only a few kilometres away, who come to fish here on a regular basis. People also come to fish here occasionally from elsewhere in Devon – Exeter, Newton Abbot and Plymouth – but also as far away as Cornwall and Bristol in the south-west, and beyond – Birmingham and London – mainly during holiday periods. Some camp out overnight in small tents beside the water's edge. Their rods are fitted with alarms and they wear monitors on their belts or in their pockets so they can be woken up at night should a fish decide to bite.

Although there are a wide range of different kinds of fish at Squabmoor it is the presence of the carp (common carp, *Cyprinus carpio*) and Crucian carp (*Carassius carassius*) that provides the main attraction for both the anglers who live in the vicinity and those who travel here to fish from further afield. The common carp is originally from central and eastern Europe but has been domesticated and reared as a food fish in ponds across Europe since the medieval period, and in the UK have become naturalized since then. As a sport fish they are highly prized, because of their size and the skill required to catch them. The UK has a thriving carp angling market with a number of specialized magazines, such as *Carpology*, books and dedicated websites such as CarpForum and Carpfishing UK. To dedicated carp specialists the carp is no ordinary coarse fish but has an elite status and importance, with special rods, bait

Figure 10.2   Sign on fishing swim

and tactics being used, and indeed the sport has its own heroes, myths and stories (e.g. Lane 2011 and Green 2014).

The local Squabmoor fishing community have an intimate knowledge of the reservoir where they fish and actively work to enhance the locality and maintain the fishing. They have built most of the swims (fishing places) next to the reservoir, shoring them up and covering them with bark chippings, providing the materials at their own expense. It is here that they pitch the tents they use for night fishing. Apart from the RM, Squabmoor fishermen are the only people allowed to camp out on the heathlands. Together with the bailiff, who also works on a voluntary basis, they police the area informally. Fishing here is mainly a male pursuit although wives and girlfriends may accompany them sometimes. The locals have individual names for their swims: 'Birches' or 'Dead Man's,' 'Sherrif's', 'Flyn's', 'Boards', 'Pines', 'Snags', etc.

These are named either after the physical characteristics of the place, such as the presence of birch trees or pines, or in relation to individual fishermen. 'Dead Man's' is named because the ashes of a dead fisherman were spread here, 'Flyn's' after another fisherman who loved this place, etc. A map on a signboard by the reservoir, produced by a local fisherman, details sixteen places.

A map by one of the fishermen is a detailed drawing of the Squabmoor reservoir showing the depths of the lake and named fishing

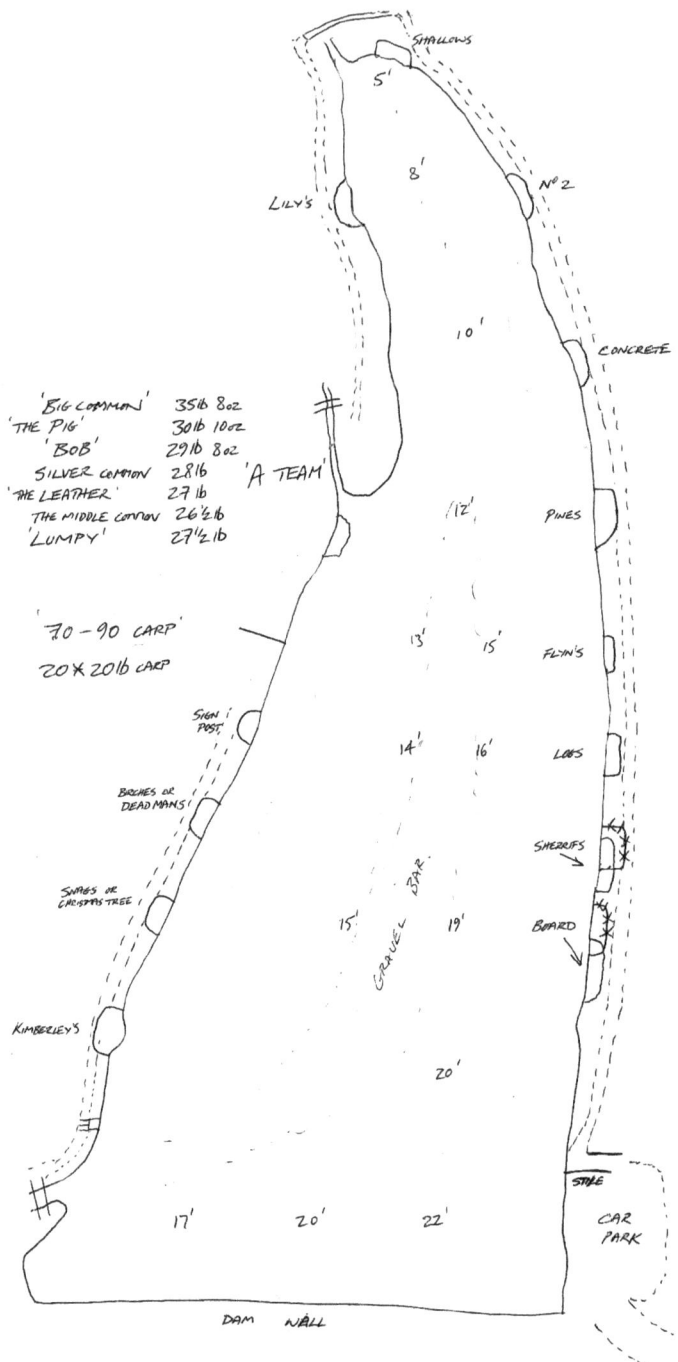

Figure 10.3 Geoff's map of Squabmoor

places, together with the car parking area and tracks running alongside it. The largest fish and their names and weights are listed on this map.

There is no best swim. Whether or not one might be successful at one swim or another is heavily dependent on the season and time of day, and fishermen tend to move around between them rather than returning to just one favoured spot. In the winter, between November and April, the fish tend to go down to the warmer bottom parts of the lake and become largely inactive. During the summer they will move around a lot more and rise to the surface. Some local fishermen also walk elsewhere on the Pebblebed heathlands occasionally but for all this is their favourite and most important place, where they spend a great deal of their free time. Some have fished here on a regular basis for twenty or thirty or more years. A passion for fishing at Squabmoor is also generational. Their fathers first took some of the contemporary anglers here to fish. The daughter of the current bailiff was even conceived here and a swim is now named after her!

The intimacy of the fishermen's knowledge of this stretch of water extends from the surrounding banks down to the depths of the reservoir. In general terms the eastern side is deeper than the western side. Running approximately down the centre is a narrow gravel bar where the water is less deep. Some, such as a fisherman named Geoff, have made detailed depth recordings of the entire bottom surface of the lake, recorded in notebooks and on plans. This is secret and personal knowledge essential for successful bottom fishing.

The affection for Squabmoor relates on the one hand to the quality of the experience of fishing that it provides and, on the other hand, to the place itself. The carp here are said to be particularly wily and difficult to catch, partly because the larger ones have been caught so many times. A fisherman can come here on many occasions and have no success. Precisely because of this the Squabmoor waters offer a real challenge and a test of the fisherman's experience and knowledge. To actually catch a carp here is then no mean achievement and directly reflects the individual prowess of the fisherman and their skill: 'It's a hard place to fish, it's not an easy place. You can sit up here weeks on end without a fish. I think I've had seven sessions, twenty-four-hour sessions, without a fish, which is good because if it were easy you wouldn't do it' (Richard, fisherman).

The place itself is equally important. Locals regard it as 'their' fishing place and are proud of the improvements they have made and the way they have maintained it. Despite the fact that people do come to fish here from beyond the immediate locality Squabmoor is still a relatively secret and unknown place in the wider UK carp fishing world. They also consider it to be a small and intimate place, unlike larger and better-known carp lakes, nestling in the midst of an AONB. Fishermen come here to be away from the crowd and for the peace and quiet the area affords. They enjoy the scenery and the wildlife that surround it: deer, squirrels, coots and moorhens, kingfishers and other water birds, dragonflies: 'I wake up and look out at this and this is my front room. What a fantastic front room … I think it's fantastic, absolutely fantastic' (Anthony). While the fishermen are here ostensibly to hook the carp, the location effectively hooks them. They enjoy the landscape as much as the fishing sitting out by the lake.

Fishing at Squabmoor combines bonds of friendship and sociability between those who come to fish here on a regular basis. There is no fishing club as such but those who come to fish here regularly behave informally like members of a club, chatting to each other on a regular basis and helping each other out. This is combined with a keen competitive edge whose principle aspect is secrecy:

> If I'm fishing and everything is going well and I'm catching fish, my bait, my rig, my location is working, I don't want to give my secrets away. I'll help anyone. But when you have something that works you don't want to give that away because the fish start getting wiser and wiser but if you keep it to yourself or just a couple of you, that tactic will work longer.
>
> (Squabmoor fisherman)

Baits used include fishmeal-based boilies, maggots, sweetcorn and halibut pellets. But which particular bait works varies according to the season, fishing technique, and the location of the swim. While some fishermen at Squabmoor purchase ready-made bait others prefer to mix their own, according to secret formulae including additives such as casein (milk protein), anchovy, liver powder, semolina, egg albumin, bird food mixes, brewer's yeast and even peppers and spices. The aim of these additives is to give off as strong as possible a food signal to the carp and make the bait irresistible to them, stimulating the really big fish into feeding. Carp have good vision, more sensitive to changing light levels than humans, effective hearing capable of detecting frequencies from 60 to 6,000 Hz, a keen sense of smell, an excellent sense of taste and can feel a hook and not take the bait. The visibility of the bait and its smell, taste and colour may all be significant considerations in bait mixes.

Bait may only be one small factor in success. Other variables include the way the bait is presented, the type and size of hook and line and the rod used, weather conditions, wind direction and water temperature. Taking all these factors into consideration carp fishing can seem akin to a form of alchemy, well suited to modern myth-making. There is no such thing as the best rig, or equipment, that can assure success. Carp tend to feed in places where they feel safe. As a consequence a 'hot spot' where there have been successful catches in the recent past may go completely cold. The fascination of carp fishing thus involves a battle of wills between the fish and the angler. It involves both material resources (rod, hook, bait type, etc.) and immaterial knowledges relating to such matters as the particular location, surface and underwater vegetation and characteristics of the lake bottom. Acute observation of surface signs may be useful: these include feeding bubbles, swirls or water movements on the lake surface. Observation of where and how other people are fishing and the tactics used will also be invaluable, as are camouflage and concealment tactics on the part of the angler. In this respect there is a certain symmetry between the RM and the Squabmoor fishermen, with the difference being that the enemy that the RM engage in their fire-fights and training exercises on Woodbury Common is fictional, while the carp are real.

The carp fishermen at Squabmoor know, more or less, how many carp live in the lake: between seventy and ninety. Of these about twenty are large carp, over 20 lbs in weight. They have no equivalent knowledge of any other species of fish. Size is paramount in carp fishing, rather than quantity: the greater the weight of the fish the better it is deemed to be. The ultimate prize is to catch the largest fish in the lake. The really big carp all have proper names and are referred to as the 'A Team'. They are the 'Big Common', weighing in at 35 lb 8 oz, 'The Pig' (30 lbs 10 oz), 'Bob' (29 lbs 8 oz), 'Silver Common' (28 lbs), 'Lumpy' (27 lbs 8oz), 'The Leather' (27 lbs) and the 'Middle Common' (26 lbs 8 oz). Geoff has caught all these over the past few years and the weights refer to the last time they were caught by him. He, like other experienced local fishermen, has lots of photographic records of the fish he has caught and a log where he records the catch, details of the location, water temperature, depth and so on. Carp can live for fifty years or more and each has its own individual characteristics, differing in terms of shape and size. 'The Pig', for example, has a slightly deformed mouth, looking somewhat pig-like. 'Popeye', one of the smaller carp, has distinctive eyes that stick right out. Unique scale patterns distinguish other fish. Local fishermen can tell exactly what fish it is because of its size and weight and the scale patterning along the body, which can vary considerably. Common carp tend to have an even and regular scale pattern whereas mirror carp (a genetic mutation) have irregular and patchy scaling making them unique and distinctive. For example, 'The D Scale' in Squabmoor has distinctive scales in the shape of the letter D. Some have a continuous line of scales along their lateral line, others may be covered in differently sized scales. Some with 'starburst' patterns have hundreds of tiny scales around the tail or belly; others may have one or more giant scales that may be clustered together. Leather carp have only a few scales.

Some of the large carp at Squabmoor are more than thirty years old. The two largest common carp originate from Cannock reservoir in Staffordshire. They were introduced here in the 1980s and have been here ever since. If an important and large carp is caught it is weighed and a photograph is taken of it held by the fisherman, witnessed by the bailiff or a friend or fellow fisherman in the vicinity to provide verification. It is then examined, treated with antiseptic cream if injured or damaged in any way with sores, leeches or cut marks, and then released. The carp have difficulty in breeding in Squabmoor but the fishermen do not at present want to run the risk of having it restocked, given the potential this has to introduce diseases or stress for the original fish. The more carp in the lake the less each fish would have to eat and they would not grow so large.

The carp in the lake with their own personal names are treated like pets. Fishermen are concerned if they do not look healthy, use cream to cure their ailments, and take pride in seeing them grow larger and larger:

> They're like my babies! Yes they are! If someone catches a fish and they've got sores on it or whatever and its down in weight it's a worrying time and everyone starts to worry … my boy is catching fish that I was catching twenty years ago. And when one of the big ones does die it is very sad, when any of them do really. It's a sad time because to me it's history. Yes, it's history. You are losing part of history.
>
> (Anthony, fisherman)

If a fish doesn't get caught for three or four years it goes on a missing list. Fish found dead are buried by the bailiff in a graveyard (unmarked) that he has established to the south of the reservoir.

The fish are not only distinctive physically. They also have their own characteristics. Some fight differently from others when hooked; some like particular areas of the lake at different times of year; some are extremely wily, others a bit dim: 'Fish are like humans: some are clever and some are stupid. And we are all different sizes aren't we? Big people and small people and some in the middle. And it's like that with fish. Some fish will never grow up to be big fish' (John, fisherman).

Conclusions

The most significant thing about fishing at Squabmoor is clearly the manner in which the biographies of the fish are linked to those of the fishermen. Their histories and lives are intertwined and this is both a personal and an emotional relationship: the fish are precious. This is only possible in the intimate arena of a small lake. Most do not go sea fishing or game fishing for trout or salmon because it requires a different 'mentality' and the relationship with the fish caught is necessarily of a different kind: it could not be personal. To catch and eat one of these carp would be quite abhorrent and some fishermen commented with disgust and incredulity that people from other ethnic groups, such as Poles, did not share the same attitude: they 'just eat them. It's not about the money. They just eat them' (Mark, fisherman). Elsewhere they said they'd been to carp-fishing waters with fences and security guards and signs on a gate saying 'No Polish or Eastern Bloc'.

The lake, through time, becomes part of the bodily memories of the fishermen, memories that are preserved in photographs of themselves and the fish that they have caught, in their fishing logs and diaries and also in photographs of the lake itself taken at different times and in different seasons: in mist, with ice, in a dappled dawn sunrise. Regular fishermen establish close personal relationships with others and a shared understanding of and relationship to the fish that they consider as being theirs in exactly the same way as a dog and its owner. Taking care of the fish and the place is a way of taking care of themselves and showing respect for non-human beings. Their embodied experience of landscape is largely static. The rod, as is the case for cyclists' bikes, becomes a sensuous extension of their embodied experience, connecting them with the fish in the water below them. The fishermen sit in their chosen swim and observe the water and its ever-changing myriad forms.

Looking out across the lake theirs is always a limited horizon seen from a low place. That which is mainly out of sight, the fish, is as important as what they can see on the surface and they must imagine the depths below them. Their contact with this watery realm is mediated by the technologies of rod and line and hook and bait that become extensions of their own bodies. Above all the Squabmoor fishermen have

Figure 10.4 Carp fisherman at Squabmoor

made this place their home. Their attachment to it differs significantly from their relationship to the rest of the heathlands and from other user groups, whose relationship with the Pebblebed landscape is both more transitory and more mobile, apart from the model aircraft flyers and their relationship with their dedicated flying field, discussed in the following chapter.

11
Model aircraft flyers: spirals and loops in the sky

'Some very strange blokes' is how the East Devon Radio Control Club describe themselves in their Company Overview on Facebook. Strange? No, but it is true that there are currently no female pilots flying model aircraft in the sky over the Pebblebed heath. The flyers' is a hobby that seems to possess a fascinating hold over its participants across the years and it is interesting to explore what is involved and the ensuing relationship with the heathland.

There has been a model airfield on the Pebblebed heaths since before the Second World War. The field moved to the present site after a sighting of a Montagu's harrier, a protected bird of prey, on Aylesbeare Common before that area was let to the RSPB as a nature reserve. Now the airstrip may be found on Woodbury Common at grid reference SY03858656, some 150 yards from the car park on the B3179, a road that traverses the heathland from east to west and borders the Woodbury and Bicton Commons. There is a slight sheltered slope leading up to a barrier and in fact you are soon confronted with four tracks to choose from. The left leads back to the B3179, the right to the next car park. The other two form a V shape: the left-hand path was originally a firebreak but now with regular use by walkers it has become an unofficial path to Woodbury Castle. The flyers joke about this: 'It's an unofficial path; it's not actually there', and amid much laughter another responds, 'That's right, you can't see that one'. Upon being asked why Clinton Devon Estates would mind it being used as a path the response is 'They try to keep people to particular directions', although it is doubtful that this is the case. The right-hand fork of the V-shaped path leads to a sign that reads: '!Warning! Model Aircraft Flying'. You know then that you have arrived. This is a very open space with views across to Woodbury Castle, the gentle hills towards Honiton and the sea that lies between Budleigh Salterton and Sidmouth. Up ahead, in the distance, is the Royal Marines' grenade range, but the woods block the view. The strip itself is about 75 × 75 metres square.

East Devon Radio Control Club (EDRCC)

There is some debate as to when the club was formed, as the original documentation has been lost, but a former chairman remembers being a founding member: 'Yes, I would say that the club itself was probably formed in the late '60s ... when there were about four or five of us that got it together, but the flying has taken place on Woodbury Common starting from before the [Second World] war'. The EDRCC is the biggest of its kind in the South-West region and one of the largest in the UK. Behind the club there is something that one flyer describes as 'a fairly tight organization'. There are over seven hundred clubs belonging to an association named the British Model Flying Association, whose mandate comes from the Royal Aero Club, the Civil Aviation Authority and Sport UK. The club also has links with the Society of Model Aeronautical Engineers and, as will be seen, the construction of the models is of great importance to the club's members. Membership cards are issued; these display the member's 'achievement level' and also show the member is insured. The 'A' level is the standard club level and this allows members to fly at public events.

There are only a few time restrictions and the flying times are 10:00 to dusk from Sunday to Friday and 10:00 to 14:00 on Saturdays and bank holidays, with a session reserved for electric models only from 14:00 to dusk on Saturdays. Most people tend to fly at weekends and those who are available on weekdays tend to fly on Tuesdays and Thursdays. A minimum of two people are required to be at the strip – one to act as a safety marshall, the other a flyer. The club has been given a licence to use the site by Clinton Devon Estate and is charged £1,250 per annum together with a third of the costs of maintaining the car park. It appears that CDE have not yet charged the club for the car park maintenance and the club itself maintains the air strip: 'We cut it, we roll it, we cut it, we fill in the holes (made by rabbits, moles and the occasional helicopter landing) and so on' (Felix, model aircraft flyer). On occasions when it is not protected adequately by the barriers, motorbikes and cars go on the strip, which the flyers say ruins it: 'You end up with skid marks and it's a couple of years before it re-grows' (Model aircraft flyer). Other occasional annoyances include cars being set alight at the entrance of their car park, but a continual nuisance is that of dog pooh: 'Just to give you an idea of the scale of the problem, G. T. used to pick up the dog mess with a shovel and put it into two whopping big heaps. Occasionally we say to people, "You are going to pick that up, aren't you?" and you get a mass of abuse'. One flyer emphasizes that he gives 'a right fight' on this issue when he's up on the Common. Some of the flyers appear to resent the charge made by CDE: 'Oh yes, they charge us to use the site whereas like you, say, horse people, dog walkers, just ordinary walkers, they're all getting it free of charge'.

Tuition is an important feature that the club provides and this usually takes place on Sunday mornings, although other times can be arranged when it is convenient for both trainee and instructor. This is described by one member as a 'buddy system':

> (There are) two transmitters, one is disabled but linked to another one that is actually flying the plane or in control of the plane and the trainer holds the real transmitter and the trainee holds the dummy. The control can be passed from transmitter to transmitter by the trainer and, just as in full size flying when they are training, you say to them, 'You have control' and they'll say, 'I have control', and then you'll fly a bit and then when the trainee gets into trouble the trainer will flick a switch and take control back.
>
> (Model aircraft flyer)

It is remarked that beginners tend to fly very high because they are afraid of the ground and they also allow the plane to go too far at times and have to be asked to turn round: 'They may have got to the point where they are no longer in control because they can't see it properly … it's very much like learning to drive a car, you do need help' (Model aircraft flyer).

The question of safety is a top priority and there are a number of facets to this. Besides the efforts of those involved in training new flyers these include awareness of one's environment when flying as well as the flight-worthiness of the model plane. The airfield marshall is also constantly on the lookout for other users of the Common and dogs. Again, just as in full-size flying, take-offs and landings can be the most dangerous aspect of this pursuit. The presence of dogs is particularly hazardous as they are often attracted to the noise of the moving aeroplane and the motion of its propeller. There are also occasions when the flyer has radio interference from an unknown source (possibly Royal Marine radio signals) and the aircraft can move in the wrong direction or even drop out of the sky: 'Planes can be very dangerous things. If you hit somebody you can easily kill or injure them' (Model aircraft flyer). On one occasion a child was hit by a plane: 'but that was the worst situation we've had'. Injury is also something that can be suffered by the flyers, who frequently injure their hands, fingers and arms: 'The top of my finger has been re-arranged (laughs)' (Model aircraft flyer). This type of injury is usually due to touching the propeller: 'You can cut a finger off with the propeller once the engine's running'. Actual air collisions are rare as there is a very strict flying area but the aeroplanes often go in the wrong direction: 'It happens every time we fly, everywhere' (Model aircraft flyer).

Other groups, particularly people coming from Cornwall and Bristol, also use the airstrip. Every year the latter group holds a 'free flight championship' on the Common. Free flight does not involve the use of radio control. As well as the use of the noisy internal combustion motor, motive power can also be accessed by quiet means such as electric, CO_2 or the traditional rubber strip motor. The plane's motors are set going and then it is released. There are set times to let the engine run, then the plane is triggered to circle – 'flight and duration is what you're aiming for and it's very competition-orientated' (Model aircraft flyer). Some of these craft have transmitters on them but without a 'bleep' they require following and tracking down to wherever they have landed.

The planes

The flyers bring their models to the strip: the means of locomotion depends on the weight of the model plane and its construction. The big, heavier planes have wheels under their carriages and can be towed. The smaller are held in the arms of the modelers, sometimes fully assembled, sometimes not. The planes then vary greatly in size and weight and the materials of construction also affect this. At this site there is a weight restriction of 10 kilograms; a typical wing span size is between 100 and 150 centimetres, but some models reach 230 centimetres or more. It would appear that a number of the new electric models have tended to be a bit smaller but these models are now increasing in size too. The technology behind these crafts has been changing: 'The electric models, you wouldn't have seen them five years ago because the technology wasn't there' (Model aircraft flyer). The flyers are aware of the noise that the non-electric aeroplanes make but a former chairman of the club believes there has only been one formal complaint about this and the flyers are adamant that the noise does not affect the birds or other wildlife. Unlike those powered by petrol or methanol, the electric models are very quiet. One flyer whose plane has a wing span of 230 centimetres describes his plane as being noisy:

> but hopefully a lot noisier than it's going to be this time next year as I've got a different silencer to put on it and it will give it more power and it will make it quieter. A tuned spike – it's something that you get on high-performance cars as well.
>
> (Model aircraft flyer)

His colleague describes this motor as being the sort of size you might find on a garden instrument such as a powerful strimmer. 'Or a motorbike', his friend interjects'; 'A little motor bike' is the rejoinder. Does size matter? No, for all of these planes have character and history as well as differences in colour, size and weight. Performance in the air is largely dependent upon the skill and experience of the flyer although it is felt that the larger models may be easier to fly as they are buffeted about less than smaller models.

The construction or building of the model planes is very much part of the pleasure for the flyers (Figure 11.1), one of whom has erected a display of model aeroplanes that he takes to shows. One flyer who has been involved in this pastime for over seventy years believes that the building of the model is more important to him than flying it: 'It is for me, yes. I mean during the war (Second World War) we used to build solid models, out of terrible timber'. Nowadays he builds scale models: 'I like things to look quite real. I build to scale, slowly, and it is very good practice at my age to have this eye–hand co-ordination. I'm not a very good flyer but I'm a keen builder and I get a lot of pleasure building models'. Some of the flyers came to the heathland with their fathers when they were young and flew models using a control-line, using the lines to fly the model in a circle before landing. Others came with chuck gliders – 'You just chucked it and see where it went … we built them and flew them, lost them and broke them' (Model aircraft flyer). Many of these fliers have been coming here for between fifty and seventy years to fly models and these occasions are remembered with affection.

One flyer, Brian, has brought two of his scaled model planes to the airstrip. One is a Vickers Wellington, the other a North American P-51 Mustang. Both of the full-size versions of these planes have huge history behind them. The Wellington's fuselage and wing structures used

Barnes Wallis's geodesic design, Wallis being the well-known inventor of the famous bouncing bomb (as in the *Dambusters* movie). Some of the older flyers can remember Wellington bombers taking off in the Second World War and Brian remarks that his plane is quieter as it is electric. His Mustang has a pretty design of a black woman sitting on one side of the plane and Brian explains its story. It transpires that during the Second World War African Americans in the American military were racially segregated from white troops, and this included the airmen. Only the Tuskegee 332nd Fighter Group was allowed to take part in overseas operations and this model of Mustang is the plane with which they were most associated. Brian's Wellington has been constructed from recycled materials such as insulation foam and plastic, which have been carved out. The Wellington's batteries have a camouflaged recycled pop bottle covering them, for example. His flying colleagues make jokes about him going round bins looking for materials. He says: 'I'm not quite as bad as that but my wife drags me away from the recycling bins. When I pass a skip she says, "don't look in that!"' However, looking at Brian's scaled aeroplanes it is clear that much skill and time is required to construct his models, the Wellington taking about two months to build.

Figure 11.1 A model aircraft enthusiast and his plane

As many of the flyers have more than one plane they are asked where the models are stored and one flyer responds: 'I've got aeroplanes in the shed, aeroplanes in the garage, aeroplanes in the cellar and aeroplanes in the house'. 'My, how many aeroplanes have you got?' asks a fellow flyer. 'Too many', he responds, and everyone laughs.

Flying

Apart from the time restrictions, weather is also the main factor affecting when flying can take place. Strong wind, rain and mist all mitigate against the flyers – rain getting into the transmitter can mean losing the link between plane and transmitter, leading the plane to crash. The distance flown is dependent on the ability to see the plane: 'As long as you can see it, if the radio control is in good order and set up correctly, you can fly as far as you can see it but any further and you will not have control because you can't see it'.

How often a plane is flown depends of course upon the individual member and this varies a great deal, from a few times a week to the occasional weekend. If a number of people are flying they tend to fly in what is referred to as circuits. Whether it is a left-hand circuit or right-hand circuit depends upon the direction of the wind. Take-offs and landings are at ninety degrees to the wind and constant awareness of the landscape is very necessary: 'Wind direction, people, air flow, making sure you're at the right height, right speed, coming in over gorse and so on, not hitting anything. Always, you are aware of your surroundings and you're aware of the model' (Model aircraft flyer). Although taking off is relatively easy compared to flying and landing, when in training the beginner only learns how to take off once they have learnt how to fly and land. 'The theory of that is the pupil isn't tempted to go and fly by themselves before learning properly because they haven't learnt to take off' (Model aircraft flyer).

One flyer, Felix, explains before taking off for a demonstration flight how he holds the controls: 'There's two schools of thought. One is thumbs on top, which is what I do, and the other way is using fingers and thumbs and is done with the control hung from a strap around the neck. I get on with thumbs much better'. He then shows us how the ailerons move and what they do. The aileron (French for 'little wing') is a small, hinged flight control surface found on each wing; these are used to control the aircraft when performing rolls for example. These ailerons are interconnected so that when one goes up the other goes down;

the downward motion increases the lift on the plane's wing with the up-going aileron reducing the wing's lift. The first thing Felix does is to start the propeller manually; he then checks which direction the wind is blowing, determining this by the airstrip's windsock. He moves in to half-throttle and pauses when people appear on the path. Finally he moves the thrust control, giving his craft momentum and creating lift on the wings. He takes off in the direction of the windsock and commences some aerobatic manoeuvres. Lines and loops, rolls, spins, figures of eight and stall turns are all performed; interestingly these are the same technical terms as for full-sized light aircraft manoeuvres, as are the terms used in describing how the manoeuvres are performed. There is an inside loop where the nose of the plane is pulled up, resulting in positive G-force, and an outside loop where the nose of the plane tilts downwards and a negative G is drawn. One flyer remarks, 'If you were in a plane and it did that you wouldn't like it. Doing the loop is uncomfortable'. In fact in full-sized aircraft doing such loops can cause a blackout for the pilot in the case of the positive G and a 'red-out' (when excessive blood is pumped into the pilot's eyes) in the negative G scenario.

When Felix decides what he wants to do he does not look at the controls but at the plane. Brain, eye and hand coordination are seamless. Felix remarks that things are very different when a flyer is inexperienced: 'You're thinking about every single move and you're thinking about what your hands are doing, you're thinking about everything. Once you've become more experienced you just do it and basically your mind goes to the plane rather than down to the controls; your fingers just do it'. The flyers agree that when they are on the airstrip they are in their own world. 'When you're actually flying you're totally focused. Take your eyes off it for a second and you could be in trouble' (Model aircraft flyer). One flyer has had sitting profile photographs taken of himself in flying gear. These are placed in his model plane's cockpit and from a distance it really does look as if he is sitting in a plane flying it. (Figure 11.2). There is some discussion as to whether, in their minds, the flyers are in the cockpit when flying, and it is decided that this is not the case: 'You've got to have distance to react to orientation as you see it' ... 'Yes, you're *with* the plane but you're on the ground, totally concentrating, handling the controls. You see where you are in relation to everywhere else, where the aeroplane is in relation to everything else and where the aeroplane is in relation to the ground'. As far as operating the controls is concerned, when the plane is flying away from the flyer left is left and right is right but when it comes towards the flyer, the opposite is the case. If the flyer wishes to bank towards the right, the left wing has to drop: 'You've got

Figure 11.2 The humanized cockpit

to think the other way round and that's one of the fundamental things that every beginner has to get' (Model aircraft flyer). This is like looking at one's moving reflection in a mirror and is very different from flying an actual light aircraft. It is remarked that pilots of light aircraft sometimes have difficulty learning to fly model planes because they are used to a joystick and not this mirror effect when using the controls.

Asking why a flyer gets pleasure in participating in this hobby invokes a thoughtful discussion: 'It's so hard to describe! Oh gosh! Um, it's next to the real thing'. It is recognized that some people may regard this as an 'anorak' kind of hobby but to flyers, when an aeroplane is actually flying, using the air to fly on, 'it's a thing of beauty'. Some flyers have always been interested in aeroplanes:

> Boys would come here with their parents and then they would start getting interested in motorbikes and girls and it packed off, dropped off. But late on in life, after you've started a family or whatever, you start to come back to it again, 'Ah yes, I think I'll have a go at that again', and back you come, you know the way it works.
>
> (Model aircraft flyer)

Another describes his enthrallment: 'Once you've been hooked on aircraft, and I've been hooked all my life, it's the sit of the aeroplane, the aeroplane size, you know, the look of the thing in the air'. Some flyers get particular pleasure in making a scale-model craft fly like a 'real aeroplane', making a Wellington fly like a Wellington, for example. There is also a sense of achievement in successfully controlling something they are not in immediate physical contact with. The hand movements on the controls are very delicate and very precise: 'You can hardly see the hands move and yet the aeroplane will respond'. Some flyers set goals or keep records such as the number of flights flown, total airborne time and public displays attended. Other flyers have enriched their experience by becoming tug pilots – joining up with a fellow flyer and aerotowing gliders. For there is also a sense of a community of flyers and this may be seen on their website, which includes a display of events, photographs of all the club's model aircraft, and an exchange of 'stories'. This communal sense sometimes extends outside flying times. The flyers become friends and get together on a social basis as well. People also come to watch the flyers. Some bring chairs so they can sit and watch the flyers perform their circuits. This is particularly true on a summer Sunday morning: one flyer remarks that some people turn up in the early afternoon just as the flyers are leaving and get quite upset that they have missed the display. Performing for other people's pleasure is part of the excitement.

The environment

The relationship the flyer has with the heathland can be twofold – one is mobile and to do with what he experiences when he comes to the heath to walk, or paint, explore its geology, its history, and study or observe nature, as these are all some of the other pursuits followed by the flyers. The other is when he comes here to fly, and this is quite singular in that it concerns one area, part of it being the airstrip, which is 'fixed' and signposted, and actually one of the few written signs in this landscape, the others being the notification signs put up by CDE about the wildlife, South West Lakes Trust about fishing at Squabmoor, and information regarding the Iron Age hill fort, Woodbury Castle. The remainder of the area is that visible to the flyer and this of course depends upon the weather at the time of each visit. A map made by one of our flyers shows the approach road to the airstrip from the west, marking landscaping mounds along it, the path to the flying field, the heath beyond, a typical loop of a plane (arrowed), and Woodbury Castle. Equal importance is given to a sketch of a plane (Figure 11.3). As has been noted, the club carefully maintains the airstrip and some flyers feel a sense of ownership over this particular area of heathland: 'I know it doesn't belong to me, I know I'm only borrowing it for the time that I'm there, but, yes, because we have access, I think, "Yes, that's our place to be"' (Model aircraft flyer). Another remarks that he moved to be near to the flying club. In fact, some cyclists have also stated choosing to live close to the heathland in order to take part in activities there.

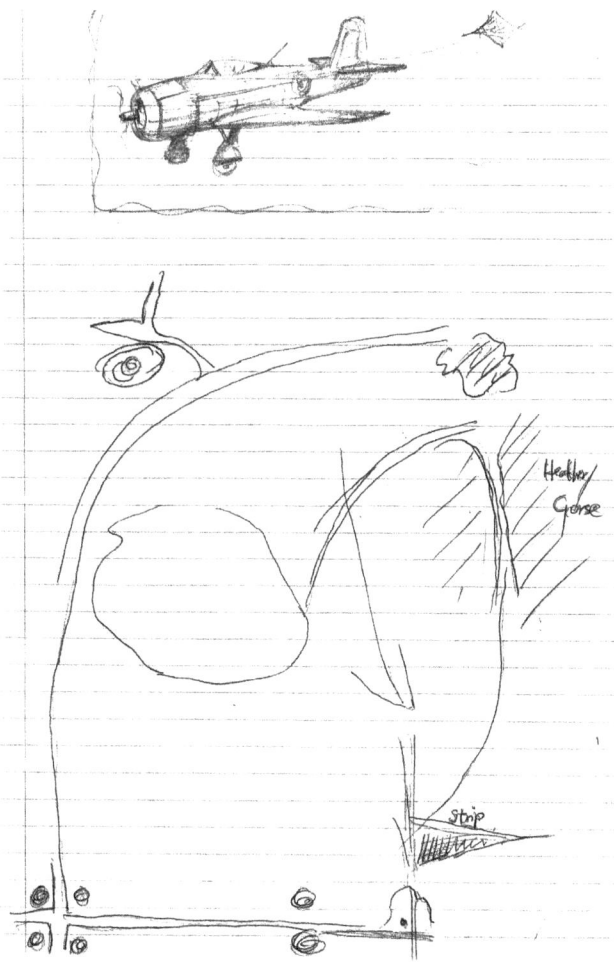

Figure 11.3 Model aircraft flyer's map

The club is very aware of the danger of heathland fires resulting from their activities and the club rules stipulate this: 'No fires of any description shall be lit on the flying site at any time'. However, there is also the danger of a fire starting as a result of an accident involving a model aircraft and in 1995 a large fire ensued when a model aeroplane's fuel ignited after crashing. One flyer speaks of a fire, believed to have been caused by a dropped cigarette, that started on the other side of the hill in the far distance one Sunday morning: 'We were all here until one o' clock and by that time it had burnt three parts of the way down so we beat a hasty retreat and by four 'o' clock it was here – it was really devastating'. They comment on the effects of this fire: 'Within weeks the birds were back, there was grass sprouting up and within eighteen months you wouldn't have known what had happened. One good thing it did was get rid of the gorse'. The flyers are particularly grateful when the estate cuts down the gorse around the airstrip, as it is painful finding and reclaiming a plane that goes down in this vegetation.

Although it is felt that the heathland is not a natural place in that it has to be maintained in order to keep it as heathland, some flyers do find it to be a wild place: 'It's very wild. I can't quite say it's original but it seems original and it can be bleak'. So, not only is the heath felt to be wild, so is the airstrip: 'Our site is very rough and rugged and you have to be quite good to use it … people come from other places and find our site rather difficult because of this but it's an absolutely gorgeous site to fly from'. This largely is to do with the fact that there are few trees in the area around the airstrip and thus there is much space for the aeroplanes to move around in. There is also much for the human eye to gaze upon: 'It's a lovely area … You go up there and you can see for miles, you know, all over the place'. One flyer remarks: 'I get up on the strip, look across to the sea and think, "Yeah, this is a good place to be"'. Even though it would be helpful to have a place to store their windsock and shovel, the flyers would rather there were no structures put in place there, nor amenities placed elsewhere on this landscape.

Many are aware of the different historical aspects the heath offers, both recent and ancient. Brian says: 'Links to the past are all around us. There's burial mounds and the tribe must have been fascinated by the heath. Absolutely amazing. Ritualistic, spectacular things'. Others speak of the Heath's military history and how it was used during the World Wars. Of course, this is still a landscape that is shared with the military, the Royal Marines. The flyers state that the Marines pay a vast amount of money to use the heathland and remark upon their relationship with the latter: 'The Marines aren't allowed on our strip, however, they do go on the site, deliberately, because it's a useful little place for them ... they landed a helicopter one summer ... we liaise with the camp ... we have had words backwards and forwards but we have good relations with the camp'. The airstrip is, in fact, essentially in the middle of an area of heathland the Royal Marines use most for training and the latter also feel they have a right to use this area. Thus there is bound to be slight conflict between the two bodies of users from time to time.

The flyers' relationship with the Pebblebed landscape tends to be static compared to other kinds of heath use. Unlike walkers, the flyers have no anticipation of roaming unknown tracks or of observing changes in nature: the flyer will drive to the car park and take the short walk to the airstrip, a familiar and set place. Unlike the heathland around the airstrip with its tangle of gorse and heather, the strip is flat, with carefully maintained and closely cropped grass. The flyers' activity has dictated a change to the landscape in this location and it is a maintained change – one that becomes fixed and almost unchanging, physically, in presence and in memory, and so the landscape is probably experienced primarily in this way. Although the character of the views seen from the airstrip may alter according to weather and season these remain the same views and whether they are of significance when the modeler is flying is questionable. It is, then, a form of confined embodiment; the flyer is in an almost fixed landscape, creating a temporary theatre in the sky through delicate movement with a focus on space, air flow and the flying model, rather than the landscape *per se*. His safety marshall is on the lookout for movement from others (walkers, dogs, birds) within this landscape, either on the ground or in the air.

Conclusions

The relationship of the flyers with the heathland is both personal and historical. Many members of the club are local to the area and several have been coming here since their childhood. Sadly, one club member, Dennis Lippet, died on the path to the airstrip and his family placed a birch tree together with an engraved memorial notice. Unfortunately the tree died. It was replaced with another but when it died too, it was not replaced and the notice was returned to the family. Thus the airstrip and the visible area around it is a place wherein lies the fusing of personal memory and historical events, whether such events are military or taking place in nature. In some cases the model aeroplane reflects other interests the flyer may have and it could be said that in this way the model takes on the personality of its owner, its producer of construction and flight. For example, the flyers' aeroplanes are sometimes replica scale models of those that once flew through the heathland sky, and these flyers take particular pleasure not only in possessing the history of the original aeroplane but also in making their model planes fly in the same way as those full-size planes of seventy years ago or more. There is embodiment between flyer, the heathland landscape and the physics of flight. Some model aircraft flyers enjoy flying slowly, others prefer speed and their type of aircraft often reflect such preferences, but for all there is an awareness of space and direction whilst being totally absorbed in creating the movement of line, spiral, loop and roll. First there is the plane visibly performing these manoeuvres, then, in the human imagination, there can almost be an invisible line, spiral, loop and roll trailing from behind the plane. For flying the model is a skilled art and the manoeuvres performed leave this faint moving calligraphy in mind and memory long after observing the performance.

12
Conclusions

Here we return to some of the key themes discussed in the introduction – materiality, embodiment, contestation and emotion – and the manner in which they are played out in the lives of persons in this particular landscape.

Back to materiality: what landscapes want and what they do

'Landscape has been considered throughout this book as a 'thing', part of a wider category – material culture. From such a perspective it is the most difficult and complex kind of thing that we might study and that, for us, provides the underlying rationale for the study. To paraphrase Marx, people work and use this landscape under circumstances that are not of their own choosing. The landscape is always already out there; its sensuous material qualities, its shapes and colours endure. But humanized cultural landscapes, such as this, want and need embodied persons, because people form part of landscapes and vice versa. Materiality resides in the pebbles and the vegetation and the earth in relation to the inhabitation of people. The fact that persons may typically take the landscapes that they use and inhabit for granted and seemingly not think about them at all makes these landscapes all the more powerful in framing and producing identities and values. People of necessity act as part of the same material world that they inhabit. We have tried to show that the material character of this landscape is fundamental in the manner in which people both experience it and how they may subsequently think through, discursively express and rationalize these experiences. A material landscape and a material mind coalesce to form part of each other. To the RM recruit the landscape that we have described may be a place of nightmares, that quite literally to use another of Marx's phrases 'weigh down on the brain of the living'. To others the same landscape is an endless source of joy and delight. But, whatever individual people think about it, this is the same material landscape.

To acknowledge that landscape can have multiple meanings and produce different emotional responses does not support a claim that they can mean anything, that landscape exists only in the mind. This is to stress that the manner in which people think is not the product of an untrammelled human mind that can think in any way it likes but is derived from embodied, perceptual, sensory and kinaesthetic experiences. Consciousness and embodied social being are always materially situated. Inevitably, different experiences of the same landscape give rise to varying personal and emotional responses. It would be bizarre if this were not the case. The material landscape is thus a highly variable resource for thought, both producing different kinds of responses in the manner in which it is encountered and constraining thought through its brute materiality. This is a position running counter to any crude form of idealism or cultural relativism in which there might be as many radically different manifestations and visions of a landscape as there are people to experience it. Landscapes are not inert; they are an active presence in which the identity of landscapes and the identities of people that inhabit them are indelibly intertwined so that they co-produce each other in dynamic ways that always change through time. Personal and social experience of a landscape is never totalizing, it is inevitably partial and from a point of view. Only an all-seeing and knowing God might be expected not to have a point of view. People take some material aspects and experiences and amplify them, inevitably ignoring others. This messy and ambiguous partiality of subjective experience is itself the outcome of a material relation.

Unnatural nature

This messiness is reflected throughout the discussions in the text of what nature is supposed to be. They emerge over and over again with different people and groups thinking through the term and understanding it differently. We have seen that NE and others manage nature in this particular case but that this is management undertaken in a cultural landscape that has been created and changed by people over thousands of years, from the Bronze Age to the present. One of us, studying archaeology and anthropology at Cambridge in the mid-1970s, was taught that the next ice age was approaching, based upon the evidence of deep-sea ocean cores. Everyone was going to freeze to death. Now, in the age of the anthropocene, we know the reverse is the case: we are all going to fry. The point of mentioning this is that nature has never been stable. Change, not stasis, is the norm and the landscape we have been considering has been constantly altered by human activities over thousands of years, but in many discussions stability is still regarded as the primary and 'natural' state. It is perilous to disrupt a world that should ideally be stable.

Today culture has become what nature is. In the process nature has lost something of its own self-identity and otherness through the passage of time. Notions of purity and danger (Douglas 1966) often loom large in the manner in which nature is thought through. The best kind of nature is of the unsullied, pure sort. In the particular case that has been considered maintaining purity involves attempts to remove species that should not be there and manipulate others so that they are of the correct proportion and state of growth. Otherwise nature, left to itself, becomes both dangerous and endangered.

To various degrees nature may still be regarded, as we have shown, as an entity that is absolutely different from culture, or simply yet another manifestation of people and their activities. What is absolutely clear, whatever position may be taken, is that this is a *valued* concept and it is required for people to act and labour in the landscape. Maintaining a concept of nature and that there is a natural world is an indispensable source of meaning and value. So nature is a concept that has practical utility. It does work in the world and facilitates the social imaginary, the dream of a better future for ourselves and our children than what we currently have. The politics of nature is a protest against the manner in which capitalist economies produce environmental crisis. Although we might dispense with nature because we realize that it does not exist, we actually require nature to perform practical work in the world. Nature is, then, the medium and outcome of political practice.

Emotion is there: we are involved and nature thus becomes therapeutic. We care, and caring for nature is also about caring for the self, finding meaning in the world for many and a reason to live. A sense of the otherness and difference of nature is absolutely essential to anyone conserving the landscape. In this respect nature is perhaps best thought of as a purely political concept. It is required primarily in visions of a better life and a better society, a political tool. The 'content' of nature, what is supposed to be in or part of it, and that which resides outside, scarcely matters to anyone other than academics. We might say it is just an intellectual game to be played. People think through the concept as they will but it needs to be there for them to be able to think about the landscape at all. To put it another way, nature and a natural world is absolutely essential to emotional well-being.

In debates about the futures of threatened landscapes nature is an indispensable part of a rhetorical armoury stimulating action, far more than being merely one side of an abstracted logocentric opposition to be textually deconstructed. Nature brings tears to people's eyes; it is something worth fighting for, a reason for living and loving. Here we can recall a personal incident as a poignant illustration of the point. Chris was talking to a group of fifty people on an open day during the excavation of a prehistoric cairn. Having talked about the pebbles from which the cairn was made and what was found he mentioned the significance of the heathland landscape for our knowledge of the past. It had not been destroyed by ploughing, unlike most of lowland Britain, but was now under threat from topsoil scraping. A lady burst into uncontrollable sobbing because for her this would inevitably mean the end of the silver-studded blue butterfly, a species that she had spent years counting and monitoring, and it required bare ground in order to live.

Put in the broadest sense nature is discursive power. It enables and empowers resistance to the corrosive forces of capital and economic development. In this sense nature is indeed 'good to think', part of a politics of identity.

Embodiment in practice

Another key theme of the book has stressed embodied performative practices in the landscape. For the most part people know how to go on without explicitly thinking about what they are doing and why in a routinized and habitual manner, whether this is the RM recruits enduring their endurance course, environmental volunteers involved in scrub clearance or walkers, cyclists or horse riders traversing the landscape in various ways. They forget about their material bodily involvement until through exhaustion and pain they are forcibly reminded of it. Participation in environmental work, or in walking, or flying a model aircraft creates a dynamic interrelationship between people and landscape through which they come to know themselves. The emotions portrayed or acted out here may be those of an inner emotionality.

By contrast our artists' emotional involvement may sometimes be highly conscious and thought through in the manner in which they respond to place. Yet the performing arts in particular provide aspects of experience that elude words, expressed through the medium of the body itself, and as such may be regarded as transformative. Phelan describes the performing body as 'metonymic of self, of character … but in the plenitude of its apparent visibility and availability, the performer disappears and represents something else' (Phelan 1993: 150). Thus embodied subjective feelings facilitate exploration of both self and other. They exemplify the manner in which shapes, forms, colour, touch and co-beings (people or fish or other animals in our discussions here) can become social and emotional agents for change and transformation. People both find themselves and lose themselves in landscapes in relation to different performative practices. Thus embodiment, like landscape, is a multifaceted concept relating mind to body and involving different types of consciousness: practical mastery or knowing how to go on in a particular material and social context or discursive expression; relationships with the earth, tools and instruments, animals and people.

A storied landscape: emotion, time, memory and place

Emotion and a feeling for place form an ontological basis for the human capacity to experience meaning. This is not something extraordinary but part of ordinary bodily experience, the means by which we touch the world and are in turn touched by it. Part of these processes of embodied knowledges involve the manner in which different individuals and groups create stories about this landscape and objectify these stories in relation to place through naming and activity, and the manner in which they approach places, directional and orientational relationships and move between them. These platial stories differ according to their material relationship with the landscape and the events that they recall. Personal biographies relate individuals to the landscape, so much so that they trace out part of their lives in terms of the places they have been and the work or activities that they have been involved with. Three of the groups we have considered – fishermen, model aircraft flyers and archaeologists excavating sites in the landscape – have a relationship to place that is primarily static; they repetitively go to the same places and may develop a particular affection for them. In this respect it is interesting to note that all these three groups have in different ways commemorated their dead in place in discreet ways: a favoured fishing spot; a plaque for a deceased

Figure 12.1 Pebble memorial to the archaeologist George Carter

flyer; personally by Chris, who scattered the ashes of Tor the dog on the excavated prehistoric cairn named after her, two years after the research project had ended.

The archaeologists also constructed a pebble memorial to George Carter, the pioneer and highly imaginative archaeologist who worked here in the 1930s.

This was in the form of a bird. Carter had suggested that he had seen the fragmentary representation of a bird on the surface of one of the prehistoric cairns that he had excavated on Woodbury Common (Carter 1936) and this and some of his other ideas had formed a background informing the excavations. The bird was made from pebbles that had been excavated from the nearby cairn, carefully selecting the colours to create a Dartford warbler, one of the key endangered species on the heathland. It was a symbolic act of engagement with the environmentalists that had resulted from a fraught relationship over conservation policy and practices with the archaeologists discussed at length in Chapter 2.

There are benches with plaques on Aylesbeare Common in the RSPB reserve, affording distant views that commemorate those who loved this place, and there is a spiritual tree near to Woodbury Castle decorated with ribbons and with offerings of flowers, materializing it as a place for memories and performances. In these and a myriad of other small fleeting practices that leave little or no trace people celebrate their personal and emotional connection with this landscape as a place for memory work. In this manner emotions become objectified and situated within the landscape. Memory and place and how people construct a sense of historicity in relation to a landscape in which pebbles are such an enduring feature is the focus of another work (Tilley in prep.)

The maps we asked people to draw of the landscape in a more formal and structured manner also materialize and name memories of both path and place. For the heathland environmental managers they depict this entire landscape in a 'platial' sense. It is the whole that is significant for them in its internal relationships. Only one RM map showed the entire heath and this was made by a senior officer, also responsible for its entire management, but from a military and strategic point of view. For others it is the part of the heath that is most familiar to them that is being depicted. For some it is the sequential relationship between places along walks taken or along a cycle or horse ride that are being shown. The maps depict journeys actually made, sedimented into memory and the ways in which places are encountered. They tell a story of encounter and experience, a mode of inhabiting the world. Together these maps situate personal, biographical and emotional attachments to place and have a visceral connection to lived experience. We have briefly commented on them in the text but in a way this is unnecessary. As a visual medium they are another way of telling requiring no words.

This is a layered landscape, layered in terms of archaeological and historical temporalities, layered in terms of places within it and tracks to follow across it. Prehistoric monuments become stable and enduring orientation points in the present while tracks shift and change in labyrinthine fashion, as do the patterns of vegetation and human association. Places in the landscape are entwined as knots of meaningful associations that are very different from mere dots on a spatial map (Ingold 2007: 101). As we have seen in different chapters people working on and using the heathlands have their own names for places. The RM name and number places significant to them, so do archaeologists creating their own names for places without them. Fishermen, environmental volunteers, horse riders, cyclists, walkers; all, as we have seen, create their own and different relationships. These are their own names and are unknown to others. They humanize this landscape, actively creating place out of space, and their memories sit in these places and are recalled as they visit or talk to others about them. Place memories are fundamental in establishing and maintaining social bonds. Each troop and generation of RM recruits can share a common experience of this landscape in their 'dits'. Like prehistoric monuments and tracks some names endure, others are lost to be replaced by others. Their presence or absence is part of an ongoing temporal dialectic of embodiment and experience.

Topophilia and topophobia

These two terms were coined by the geographer Yi-Fu Tuan (1974). By topophilia he refers to a love or an affection for place, somewhere we feel secure and at home. Topophobia is the reverse, a fear or a loathing of places in which we feel insecure. The archaeologists worked in three places in the east, north and on the west of the heathlands. Everyone, without exception, enjoyed the experience of working on Colaton Raleigh Common. Members of the team developed a particular emotional affection for the prehistoric cairns they excavated there. Here the team was high up and the location afforded sweeping views across the heath and sea. The view, the ever-changing weather and the patterns of the clouds were constant topics of conversation. Nobody enjoyed the experience of working at Jacob's Well, a place located in a bog in the midst of a gloomy pine plantation. The discovery of a dead and mutilated bird on the walk and very near to the site one morning led one female member of the excavation team to remark that this was a bad place. People who had worked previously on Colaton Raleigh Common high up in the open landscape preferred being there, enjoyed the experience more, and had developed a particular affection for that place. Indeed its memory was so strong that being anywhere else on the heathland afterwards became a lesser experience.

This is one of countless other little anecdotes that could be told about this landscape, illustrating the manner in which place memories and emotions are deeply embedded in sensory experience. A companion volume to this might consist solely of those stories. All those who work in or use the heathland have their own place memories, their own stories and their own ways to relate to and socially construct this landscape. Inevitably for some they are more extensive than others and become related to the heathland itself rather than particular places within it. For heathland managers the heath itself was platial in character. In other words they had no particular place preferences within it. As we have discussed, all RM recruits hate this heathland where they have undergone suffering on countless occasions, but if they return to it as trainers their attitude may change; it becomes a beautiful place. For some, particularly artists and poets, this is a deeply spiritual landscape with mystical powers, to others it will always be a harsh, forbidding and cruel place. Memory sits in bodies in places in which past and present coalesce and find a new unity. Returning to place, experiencing its material presence plays an active role in recollection in a profound sense, linking persons and events, situating them in a landscape.

Contestation: an ordinary landscape?

The analysis presented in this book has tried to demonstrate the manner in which the particular landscape being considered is fractured, mutable, always in the process of becoming in which change rather than stasis is the norm. Different individuals and groups think about and engage with the landscape in radically different ways and this is significantly related to their activity and involvement. As a consequence landscape is contested either explicitly or implicitly. People may actively protest about the actions of others or, more usually, feel more or less resentful but keep these thoughts to themselves, just putting up with those things that they can do nothing about. Conflict is normal.

Some might argue that the study being presented is abnormal – most landscapes are not like this and the case in hand has been selected in a tautologous way to highlight precisely these aspects. When we began this research this was a landscape about which we knew very little. Its selection for an anthropological study was not based on any supposed potential for revealing conflict and contestation. The project was undertaken to run in tandem with an archaeological and historical study of the same landscape that was starting and for which it *had* been pre-selected as having considerable potential. Our research revealed, rather than set out to investigate, what was already there, and there is nothing that is either unusual or extreme about this particular case. Scratch the surface, we would argue, and all landscapes are like this. The details will inevitably differ but the widespread notion of harmony and stability in landscape use, meaning, value and perception only has any relevance today as a myth. In many respects this is an ordinary and mundane landscape differing little from others, except perhaps in relation to its particular geology, vegetation and history. People make of it what they will; indeed conflict is a vital part of what landscapes are and in a very real sense this gives landscapes their vitality and makes them into a living presence, something that matters, and gives them dynamism and emotional presence and value in people's lives. A harmonious landscape would be one that is socially dead, meaningless and irrelevant: conflict both engages and empowers in profound ways.

These conflicts over landscape and its meaning and significance are never likely to be about single issues such as whether or not a quarry development or that of a wind farm should go ahead, although they are typically considered and presented as such. They are always multiple and far more complex than that, opaque, stubborn and disparate mediums for thought and action. Some aspects of contestation are visible at the surface; others underlie them in layers within. Heterogeneity is the norm with landscape the palimpsest. Conflict resolution is not a kind of process with harmony being the outcome. Typically it involves compromise and muddling through, accepting in one way or another what others do, arriving at some form of at least tacit consensus, as with the relationships between environmentalists and the RM or between archaeologists and environmentalists – or the horse riders and walkers and cyclists discussed in the book, carrying on and muddling through. We might even say that this is a 'very British' compromise. Inevitably some issues are never resolved and persist; others, such as the legitimacy of a new quarry development, discussed in Chapter 5, fall into the background with the passage of time.

There is another sense of ordinariness that we wish to discuss that is highly relevant and important to a study of landscape. Throughout the book we have discussed a series of ordinary practices. There is nothing particularly unusual about people walking in a landscape, horse riding, fishing, cutting down a gorse bush or flying a model aircraft. These are all aspects of contemporary culture, taken for granted, rarely examined, seemingly perhaps not worth studying or taken as serious objects of study. But everywhere that we look the everyday and the ordinary become extraordinary. There is a plurality of different material practices and material worlds at play, from the manner in which a bike is ridden or the gear worn to the naming of fishing places, to the manner in which someone walks and relates to a dog. We find not homogeneity but endless diversity, flows of meaning and significance in situated small acts.

This we would argue is the locus of our contemporary culture. Look at a fisherman and you find a whole social and symbolic world in a relation between rod and lake. The ordinary is not a superficial manifestation of culture. It only presents itself as such and hides its enormous depth and complexity if we do not take it seriously. Start to investigate the surface and examine people, their practices and the materiality of the everyday and a new world is revealed, a lived world in which experience and knowledge is embodied in the practices of people in relation to others and things. Grand theories such as Marxist perspectives on the social provide a depth ontology, as do structuralist perspectives but in a radically different manner. In both the mantra becomes: ignore the superficiality of everyday life. Dig deeper and you will find what is really going on – depth structures that generate the everyday that can be happily ignored as trivial, a theoretical tradition carried on in the writings of Bourdieu (1977), Giddens (1984) and others.

By contrast the broadly phenomenological perspective taken in this book aims to show that such a view of culture and society is fundamentally misguided. Depth, what really matters, does not reside deep down, underpinning or providing a foundation for culture. It resides within the surface and is everywhere around us. So the project of analysis becomes the recognition and the bringing forth to consciousness of the extraordinary character of the ordinary. That is another kind of grand project worth undertaking and here we have, no doubt, only been able to undertake it in a rudimentary manner. The methodology for doing this is simple and followed by all anthropologists. We attempt to understand this world through the process of immersing our embodied selves in it and participating in it. Our body, then, is our primary research tool. We are in that sense always part of and in the study. Whether acknowledged or not, all anthropological research is thus phenomenological research. Research becomes not an abstracted practice of applying external ideas and seeking generalities (sometimes strangely described as being objective) but arises from and is grounded in the study itself. Social and cultural anthropology as a discipline with grand pretensions to knowledge has always valorized discussions of social and political structures, attempted to unravel the intricacies of rituals and cosmologies and myth through its depth models. In its relative and continuing neglect of the humdrum material world in which people actually live, we might suggest, it has often been misguided about both its objects and subjects of study.

Another concomitant perspective that arises from a phenomenological intellectual tradition concerns the manner of representation: the manner in which research and its results get written into texts. The normative anthropological view is that there is a lower-level kind of activity that we term mere description and a higher level kind of activity called analysis. The two are separate and the former should inexorably lead to the latter. From the broadly phenomenological tradition on which we draw the two cannot be so neatly separated. The analysis is *in* the description. Observation and description are *themselves* social acts, part and parcel of a reflective and subjective creation of meaning and significance. They are both highly selective in that we cannot describe and observe everything and what we do describe and observe depends on what we think might be important, and is always from a point of view in both a bodily and theoretical sense. So we always pre-frame our studies while, of course, trying to remain open to being surprised and in the process re-making the frame. In conclusions such as this the normative expectation is that the generalities, or in other words a series of decontextualized abstractions, will be brought out of previous descriptions, the plot or story of the book will become unravelled and presented as the fundamental essence of the rest. This is a perspective that we wish to have at least partially avoided here.

Landscape provides a powerful medium for anthropological thought not because we can pin it down and define its study, or indeed define what it is supposed to be – this or that, or something other. Its significance derives from it being a dynamic, holistic, material presence through which we can creatively think people's social worlds, using the medium of the material world that they inhabit. That is the project of an anthropology of landscape. In lieu of further words we present a collage of images embodying personal memories.

Figure 12.2 Memory collage

References

Abramson, A. and R. Fletcher. (2007). 'Recreating the vertical: rock-climbing as epic and deep eco-play'. *Anthropology Today* 23 (6): 3–7.

Agrawal, A. (2005). *Environmentality: Technologies of Government and the Making of Subjects*, Durham: Duke University Press.

Agrawal, A. and C. C. Gibson (1999). 'Enchantment and disenchantment: the role of community in natural resource conservation'. *World Development* 27 (4): 629–49.

Agrawal, A. and K. Redford (2009). 'Conservation and displacement: an overview', *Conservation and Society* 7 (1): 1–10.

Anderson, B. (1991). *Imagined Communities*, (rev. edition). London: Verso.

Appadurai, A. (1996). 'The production of locality' in A. Appadurai *Modernity at Large*, Minneapolis: University of Minnesota Press: 178–200.

Argent, G. (2012). 'Toward a privileging of nonverbal communication, corporeal synchrony and transcendence in humans and horses' in J. Smith and P. Mitchell (eds.) *Experiencing Animal Minds: An Anthology of Horse-Human Encounters*, New York: Columbia University Press, 111–28.

Arkin, L. and Dobrofsky, L. (1978). 'Military socialization and masculinity', *Journal of Social Issues* 34 (1): 151–69.

Armitage, P. (2006). *Report on the fieldtrip to Ushkan-shipiss, October 14, 2006. Report to Innu Nation*. Submitted to the Joint Review Panel for the Environmental Assessment of the Lower Churchill Hydroelectric Generation Project. www.kuekuatsheu.ca

—— (2007). *Innu Kaishitshissenitak Mishta-shipu: Innu Environmental Knowledge of the Mishta-shipu (Churchill River) Area of Labrador in Relation to the Proposed Lower Churchill Project. Report of the work of the Innu Traditional Knowledge Committee to the joint Innu Nation – Newfoundland and Labrador Hydro Task Force*. Submitted to the Joint Review Panel for the Environmental Assessment of the Lower Churchill Hydroelectric Generation Project. www.kuekuatsheu.ca

—— (2010). *Innu of Labrador Contemporary Land Use Study. Report to Innu Nation (Sheshatshiu and Natuashish, Labrador)*. Submitted to the Joint Review Panel for the Environmental Assessment of the Lower Churchill Hydroelectric Generation Project. www.kuekuatsheu.ca

—— (2011). *An Assessment of Lower Churchill Project Effects on Labrador Innu Land Use and Occupancy. Report to Innu Nation (Sheshatshiu and Natuashish, Labrador)*. Submitted to the Joint Review Panel for the Environmental

Assessment of the Lower Churchill Hydroelectric Generation Project. www.kuekuatsheu.ca

Armitage, P. and M. Stopp (2003). *Labrador Innu Land Use in Relation to the Proposed Trans Labrador Highway, Cartwright Junction to Happy Valley-Goose Bay, and Assessment of Highway Effects on Innu Land Use*. Report to Department of Works, Services and Transportation, Government of Newfoundland and Labrador. www.kuekuatsheu.ca

Arnason, A., N. Ellison, J. Vergunst and A. Whitehouse (eds.) (2012). *Landscapes Beyond Land: Routes, Aesthetics, Narratives*, Oxford: Berghahn.

Barrett, F. (1996). 'The organizational construction of hegemonic masculinity: The case of the U.S. navy'. *Gender, Work and Organization* 3 (3): 129–42.

Bass, S. (2001). Change Towards Sustainability in Resource Use: Lessons from the Forest Sector. Report commissioned by the Mining, Minerals and Sustainable Development project of the International Institute for Environment and Development (IIED). www.iied.org.

Basso, K. (1984). '"Stalking with stories": names, places and moral narratives among the Western Apache' in E. Bruner (ed.) *Text, Play and Story*, Prospect Heights: Waveland Press.

—— (1996) 'Wisdom sits in places: notes on a Western Apache landscape' in S. Feld and K. H. Basso (eds.) *Senses of Place*, Santa Fe: School of American Research, 53–90.

Basurto, X. (2013). 'Bureaucratic barriers limit local participatory governance in protected areas in Costa Rica'. *Conservation and Society* 11 (1): 16–28.

BBC News (2010). 'Opponents fight Devon sand quarrying'. 31 August, www.bbc.co.uk/news/uk-england-devon-11145141.

Becker, J., A. Runer, D. Neunhäuse, N. Frick, H. Resch and P. Moroder (2013). 'A prospective study of downhill mountain biking injuries'. *British Journal of Sports Medicine* 47 (7): 458–62.

Bender, B. (ed.) (1993). *Landscape: Politics and Perspectives*, Oxford: Berg.

—— (1998). *Stonehenge: Making Space*, Oxford: Berg.

—— (2006). 'Place and landscape' in C. Tilley *et al.* (eds.) *Handbook of Material Culture*: 303–14.

Bender, B. and M. Winer (eds.) (2001). *Contested Landscapes. Movement, Exile and Place*, Oxford: Berg.

Berry, W. (1987). *Home Economics*, Berkeley: Counterpoint.

Besnier, N. and S. Brownell. (2012). 'Sport, modernity and the body'. *Annual Review of Anthropology* 41: 443–459.

Bicycle Helmet Research Foundation. (n.d.) www.cyclehelmets.org

Biersack, A. (1999). Introduction: 'From the "New Ecology" to the New Ecologies'. *American Anthropologist* 101 (1): 5–18.

Bijker, W., T. Hughes and T. Pinch (2012). *The Social Construction of Technological Systems*, Cambridge, MA: MIT Press.

Birke, L. (2009). 'Interwoven lives: understanding human animal connections' in T. Holmberg (ed.) *Investigating Human/Animal Relations in Science, Culture and Work*, Uppsala: Uppsala Universitet, Centrum för Genumsvetenskap, 18–33.

Bokdam, J. and J. Gleichmann (2000). 'Effects of grazing by free ranging cattle on vegetation dynamics in a continental north-west European heathland', *Journal of Applied Ecology* 37 (3): 415–31.

Borden, I. (2001). *Skateboarding, Space and the City: Architecture and the Body*, Oxford: Berg.

Bourdieu, P. (1977). *Outline of a Theory of Practice*, Cambridge: Cambridge University Press.

—— (1984). *Distinction: A Social Critique of the Judgement of Taste*, London: Routledge.

Bramston, P., G. Pretty and C. Zammit (2011). 'Assessing environmental stewardship motivation', *Environment and Behavior* 43 (6): 776–88.

Brandon, K., L. J. Gorenflo, A. S. L. Rodrigues and R. W. Waller (2005). 'Reconciling biodiversity conservation, people, protected areas, and agricultural suitability in Mexico', *World Development* 33 (9): 1403–18.

Brandt, K. (2006). 'Intelligent bodies: embodied subjectivity in human-horse communication' in D. Waskul and P. Vannini (eds.) *Body/embodiment. Symbolic Interaction and the Sociology of the Body*, Farnham: Ashgate Publishing, 141–52.

Bray, D. B. (2007). 'From displacement conservation to place-based conservation' in K. H. Redford and E. Fearn (eds.) *Protected Areas and Human Displacement: A Conservation Perspective*, 103–5.

Brechin, S. R., P. R. Wilshusen, C. L. Fortwangler and P. C. West (2002). 'Beyond the square wheel: toward a more comprehensive understanding of biodiversity conservation as social and political process', *Society and Natural Resources* 15: 41–64.

Brighouse, U. W. (1981). *Woodbury: A View from the Beacon*, Callington: Woodbury News.

Brockington, D. and J. Igoe (2006). 'Eviction for conservation: a global overview', *Conservation and Society* 4 (3): 424–70.

Brody, H. (1982). *Maps and Dreams*, London: Faber and Faber.

Brosius, J. P. (1997). 'Endangered forest, endangered people: Environmentalist representations of indigenous knowledge', *Human Ecology* 25 (1): 47–69.

—— (1999). 'Analyses and interventions: anthropological engagements with environmentalism', *Current Anthropology* 40 (3): 277–310.

Brown, T., F. McEvoy and J. Ward (2011). 'Aggregates in England – Economic contribution and environmental cost of indigenous supply', *Resources Policy* 36 (4): 295–303.

Buchli, V. (ed.) (2002). *The Material Culture Reader*, Oxford: Berg.

Burr, J. F., C. T. Drury, A. C. Ivey, and D. E. R. Warburton (2012). 'Physiological demands of downhill mountain biking'. *Journal of Sports Sciences* 30 (16): 1777–85.

Butler, J. (1980). *Gender Trouble: Feminism and the Subversion of Identity*, London: Routledge.

Cameron-Daum, K. (2008). 'Paths of desire and well-being: an exploration of why individuals enjoy walking as a leisure activity in Southern England'. Unpublished dissertation (MRes), UCL.

Cándida-Smith, R. (2003). 'Circuits of subjectivity: Oral history and the art object'. Online. www.bancroft.berkeley.edu/ROHO/education/docs/circuits_revised.doc

Carlson, J. (n.d.). 'Gregg Chadwick and painting time'. *Fine Art Connoisseur*. Online. http://www.fineartconnoisseur.com/pages/15732772.php?

Carrier, J. (2003). 'Biography, ecology, political economy: seascape and conflict in Jamaica' in P. Stewart and A. Strathern (eds.) *Landscape, Memory and History: Anthropological Perspectives*, London: Pluto Press, 201–28.

Carrier, J. and D. Miller (1998). *Virtualism: A New Political Economy*, Oxford, UK: Berg. Cited by West, P., J. Igoe, and D. Brockington (2006). 'Parks and peoples: the social impact of protected areas', *Annual Review of Anthropology* 35: 251–77.

Carter, G. (1936). 'Unreported mounds on Woodbury Common', *Proceedings of the Devon Archaeological Exploration Society* 2: 283–94.

Carter, P. (2010). *Ground Truthing: Explorations in a Creative Region*, Crawley: University of Western Australia Publishing.

Castree, N. and B. Braun (eds.) (2001). *Social Nature: Theory, Practice and Politics*, Oxford: Blackwell, 84–111.

Cater, C. and P. Cloke (2007). 'Bodies in action: the performativity of adventure tourism'. *Anthropology Today* 23 (6): 13–16.

Chadwick, G. (2015). Discussion on emotions, walking and inspiration. [email] (Personal communication October 13–16 2015).

Classen, C. (1993). *Worlds of Sense: Exploring the Senses in History and Across Cultures*, London: Routledge.

—— (2005) *The Book of Touch*, Oxford: Berg.

Clifford, J. (1988). *The Predicament of Culture: Twentieth Century Ethnography, Literature and Art*, Cambridge, MA: Harvard University Press.

Clifford, J. and G. Marcus (eds.) (1986). *Writing Culture: The Poetics and Politics of Ethnography*, Berkeley: University of California Press.

Clinton Devon Estates (n.d.). 'Forestry and timber'. Online. www.clintondevon.com/what-we-do/forestry-and-timber.ashx

Clothier, P. (2013). *The Time Between: Paintings by Gregg Chadwick*. San Francisco: Sandra Lee Gallery.

Coessens, K. and A. Douglas (2011). *On Calendar Variations*. Banchory, Scotland: Woodend Barn.

Cohen, A. (1982). *Belonging: Identity and Social Organisation in British Rural Cultures*, Manchester: Manchester University Press.

—— (1987). *Whalsay: Symbol, Segment and Boundary in a Shetland Island Community*, Manchester: Manchester University Press.

Cohen, E. (2011). 'Once we put our helmets on, there are no more friends: The "fights" session in the Israeli army course for close-combat instructors', *Armed Forces and Society* 37 (3): 512–33.

Colchester, M. (2003) [1994]. *Salvaging Nature: Indigenous Peoples, Protected Areas and Biodiversity Conservation*, Uruguay: World Rainforest Movement, Moreton-in-Marsh: Forest Peoples Programme.

—— (2004). 'Conservation policy and indigenous peoples'. *Environmental Science and Policy* 7: 145–53.

Colchester, M. and S. Chao (2011). 'Oil palm expansion in South East Asia: an overview' in M. Colchester and S. Chao (eds.) with J. Dallinger, H. E. P. Sokhannaro, V. T. Dan and J. Villanueva, *Oil Palm Expansion in South East Asia: Trends and Implications for Local Communities and Indigenous Peoples*, Moreton-in-Marsh UK: Forest Peoples Programme, West Java Indonesia: Perkumpulan Sawit Watch, 8.

Colchester, M., N. Jiwan, M. Andiko , A. Sirait , Y. Firdaus, A. Surambo and H. Pane (2006). *Promised Land: Palm Oil and Land Acquisition in Indonesia – Implications for Local Communities and Indigenous Peoples*, Moreton-in-Marsh UK: Forest Peoples Programme, West Java Indonesia: Perkumpulan Sawit Watch.

Collier, M (2013). 'On ways of walking and making art: a personal reflection', in (eds.) C. Morrison-Bell, M. Collier, with assistance from J. Ross, *Walk On: From Richard Long to Janet Cardiff – 40 Years of Art Walking*, Art Editions North, Manchester: Cornerhouse Publications, 81–9 and http://walk.uk.net/portfolio/walk-on/

Collins, P. and A. Gallinat (eds.) (2010). *The Ethnographic Self as Resource*, New York: Berghahn.

Conley, J. M. and C. A. Williams (2008). *The Corporate Social Responsibility Movement as an Ethnographic Problem*. Paper prepared for Wenner-Gren Foundation Seminar on "Corporate Lives", School for Advanced Research, Santa Fe, NM. UNC Legal Studies Research Paper No. 1285631. Available at SSRN: http://ssrn.com/abstract=1285631 or http://dx.doi.org/10.2139/ssrn.1285631

Cooper, A. (2007). *East Devon Pebblebed Heaths*, Budleigh Salterton: Pebblebed Heaths Conservation Trust.

Cosgrove, D. and Daniels, S. (1988) *The Iconography of Landscape*, Cambridge: Cambridge University Press.

Coumans, C. (2010). 'Alternative accountability mechanisms and mining: the problems of effective impunity, human rights, and agency'. *Canadian Journal of Development Studies* 30 (1–2): 27–48.

Coupaye, L. (2013). *Growing Artefacts, Displaying Relationships*, Oxford: Berghahn.

Cox, P. (2015). 'Cycling cultures and social theory' in P. Cox (ed.) *Cycling Cultures*, Chester: University of Chester Press.

Cristache, M. (2005). 'Eco-veille network (réseau Eco-veille®): an example of action in favour of walking and of sustainable development'. Paper presented at Walk21-VI, Everyday Walking Culture, the 6th International Conference on Walking in the 21st Century, Zurich, Switzerland, 22–23 September 2005. www.walk21.ch

Csikszentmihaly, M. (1990). *Flow: The Psychology of Optimal Experience*, New York: Harper Perennial.

Cszordas, T. (1990). 'Embodiment as a paradigm for anthropology', *Ethos* 18 (1): 5–47.

Curtis, M. (2007). *Anglo American: the Alternative Report*. www.waronwant.org/campaigns/corporations-and-conflict/mining-conflict-and-abuse.

Daniels, S. and S. Rycroft (1993). 'Mapping the modern city: Alan Sillitoe's Nottingham novels', *Transactions of the Institute of British Geographers* NS18: 460–80.

Dant, T. (1998). 'Playing with things: objects and subjects in windsurfing', *Journal of Material Culture* 3 (1): 77–95.

Dant, T. and B. Wheaton (2007). 'Windsurfing: an extreme form of material and embodied interaction?', *Anthropology Today* 23 (6): 8–12.

Darrier, E. (ed.) (1999). *Discourses of the Environment*, Oxford: Blackwell.

Davis, D. and A. Maurstad (eds.) (2016). *The Meaning of Horses: Biosocial Encounters*, London: Routledge.

Davies, C. (2008). *Reflexive Ethnography*, London: Routledge.

de Koning, F., M. Aguiñaga, M. Bravo, M. Chiu, M. Lascano, T. Lozada and L. Suarez (2011). 'Bridging the gap between forest conservation and poverty alleviation: the Ecuadorian Socio Bosque program', *Environmental Science and Policy* 14 (5): 531–42.

DeMello, M. (2012). *Animals and Society: An Introduction to Human-Animal Studies*, New York: Columbia University Press.

De Nardi, S. (2014). ' "No one had asked me about that before": a focus on the body and 'other' Resistance experiences in Italian Second World War storytelling', *Oral History* 14 (Spring): 73–83.

—— (2014a). 'An embodied approach to Second World War storytelling mementoes: probing beyond the archival into the corporeality of memories of the resistance', *Journal of Material Culture* 19 (4): 443–64.

Defries, R., A. Hansen, B. L. Turner, R. Reid and J. Liu (2007). 'Land use change around Protected Areas: Management to balance human needs and ecological function'. *Ecological Society of America* 17 (4): 1031–8.

Delderfield, E. R. (1948). *Exmouth*, Exmouth: Raleigh Press.

Deleuze, G. (1992). 'Mediators', in J. Crary and S. Kwinter (eds.) *Incorporations*. NY: Urzone, 281–93. Cited by P. Laviolette (2007). 'Hazardous sport?', *Anthropology Today* 23 (6): 1.

—— (2006) [1993]. *The Fold: Leibniz and the Baroque* (trans. T. Conley), London: Continuum.

Deleuze, G. and F. Guattari (1988). *A Thousand Plateaus: Capitalism and Schizophrenia*, London: Athlone Press.

Descola, P. (2013). *Beyond Nature and Culture*, Chicago: University of Chicago Press.

Descola, P. and G. Palsson (eds.) (1996). *Nature and Society*, London: Routledge.

Desjarlais, R. and Throop, C. (2011). 'Phenomenological approaches in anthropology', *Annual Review of Anthropology* 40: 87–102.

Devon BAP (2009). Devon Biodiversity and Geodiversity Action Plan. Online. www.devon.gov.uk/dbap-birds-nightjar.pdf

Dillon, A. (n.d.). Oxfordshire-based landscape painter Anna Dillon. Online. www.annadillon.com

Dillon, B. (2007). 'Airlocked', in S. Bode, J. Millar and N. Ernst (eds.) *Waterlog: Journeys Around an Exhibition*, London: Film and Video Umbrella, 16–24.

Dodwell, E. R., B. K. Kwon, B. Hughes, D. Koo, A. Townson, A. Aludino, R. K. Simons, C. G. Fisher, M. F. Dvorak, and V. K. Noonan (2010). 'Spinal column and spinal cord injuries in mountain bikers'. *The American Journal of Sports Medicine* 38 (8): 1647–52.

Donald, B. (1997). 'Fostering volunteerism in an Environmental Stewardship Group: a report on the Task Force to Bring Back the Don, Toronto, Canada', *Journal of Environmental Planning and Management*, 40 (4): 483–505.

Douglas, M. (1966). *Purity and Danger*, London: Routledge and Kegan Paul.

—— (2003). 'The gender of the trout', *RES: Anthropology and Aesthetics* 44: 171–80.

Douny, L. (2011). 'Silk-embroidered garments as transformative processes: layering, inscribing and displaying Hausa material identities', *Journal of Material Culture* 16 (4): 401–15.

Downs, R. and D. Stea (1973). *Image and Environment: Cognitive Mapping and Spatial Behaviour*, Chicago: Aldine.

Dudley, N. (ed.) (2008). *Guidelines for Applying Protected Areas Management Categories*, Gland, Switzerland: IUCN, 8–9.

Dufrenne, M. (1973) [1953]. *The Phenomenology of Aesthetic Experience* (trans. E. S. Casey, A. A. Anderson, W. Domingo and L. Jacobson, Evanston: Northwestern University Press.

Duncan, J. (1990). *The City as Text: The Politics of Landscape Interpretation in the Kandyan Kingdom*, Cambridge: Cambridge University Press.

Dutcher, D. D., J. C. Finley, A. E. Luloff and J. B. Johnson (2007). 'Connectivity with nature as a measure of environmental values', *Environment and Behavior* 39 (4): 474–93.

Dyck, N. (2004). 'Getting into the game: anthropological perspectives on sport'. *Anthropologica* 46 (1): 3–8.

Edwards, E. (2002). 'Material beings: objecthood and ethnographic photographs', *Visual Studies* 17 (1): 67–75.

Eliot, T. S. (1950). *Selected Essays*, New York: Harcourt, Brace and Company.

Elliott, S. and R. Elliott (1994). *Woodbury Parish 1894–1994: Living a Century of Change*. (manuscript) Westcountry Studies Library – [sB/WOO 1/1894/WOO].

Elliott, S. and S. Wickenden (n.d.). *Woodbury Tree Survey*; *Woodbury Flower Study*. Online. Available via www.woodburydevon.co.uk

Eriksen, T. (2010). *Small Places, Large Issues* (3rd edition), London: Pluto Press.

Evans, T. (2007). 'The WCS Cambodia Program in the Seima Biodiversity Conservation Area', in K. Redford and E. Fearn (eds.) *Protected Areas and Human Displacement: A Conservation Perspective*, New York: Wildlife Conservation Society, 42–7.

Exeter Archaeology (2003). *East Devon Heathlands Archaeological Survey. Part 1. Report 03.26 to the East Devon Coast and Countryside Service*, Exeter: Exeter Archaeology.

Feld, S. and K. Basso (eds.) (1996). *Senses of Place*, Santa Fe: School of American Research Press.

Foster, R. (2010). 'Corporate oxymorons and the anthropology of corporations', *Dialectical Anthropology* 34 (1): 95–102.

Foucault, M. (1977). *Discipline and Punish*, New York: Vintage.

—— (1978). *The History of Sexuality*: Volume I, Harmondsworth: Penguin.

—— (1980). *Power/Knowledge*, Hassocks: Harvester.

—— (1986) [1984/1967]. 'Of other spaces' (trans. J. Miskowiec), *Diacritics* 16 (1): 22–7.

Fowler, C. (2011). 'Performing Pisgah: endurance mountain bikers generating the national forest', *Anthropology News* 52 (3): 11.

Fox, K. (2004). *Watching the English*, London: Hodder and Stoughton.

Frank, A. (1991). 'For a sociology of the body: an analytical review' in M. Featherstone and B. Turner (eds.) *The Body*, London: Sage, 36–102.

Friedel, H. and C. Tacke, S. Salzmann, B. Schulz, H. Pollig and G. Fischer (1991). Introduction to Lang, N. *Nunga und Goonya*. München: Kunstraum München, 6–9.

Frynas, J. G. (2005). 'The false developmental promise of corporate social responsibility: evidence from multinational oil companies', *International Affairs* 81 (3): 581–98.

Fulton, H. (1995). 'Into a walk into nature', in Fulton, H. *Thirty-One Horizons*. München: Lenbachhaus.

—— (2015). *Indoors Outside*, 5ª Exposición, Galería Visor, http://espaivisor.com/exposicion/indoors-outside/

Gaffin, D. (1993). 'Landscape, culture and personhood; names of places and people in the Faeroe islands', *Ethnos* 1–2: 53–72.

—— (1996). *In Place: Spatial and Social Order in a Faeroe Islands Community*, Prospect Heights: Waveland Press.

Geertz, C. (1963). *Agricultural Involution: The Processes of Ecological Change in Indonesia*. Berkeley: University of California Press. Cited by Scoones, I. (1999). 'New ecology and the social sciences: what prospects for a fruitful engagement?', *Annual Review of Anthropology* 28: 479–507.

Geismar, H. and R. Empson (2004). 'Fieldworks: a review', *Cambridge Anthropology* 24 (1): 39–50.

Gell, A. (1998). *Art and Agency: an Anthropological Theory*, Oxford: Oxford University Press.

Gilbert, M. (2014). 'Young equestrians: the horse stable as a cultural space' in J. Gillett and M. Gilbert (eds.) *Sport, Animals and Society*, London: Routledge, 233–50.

Goldsworthy, A. (1998). *Stone*, New York: Harry N. Abrams Inc.
Gooch, P. (2008). 'Feet following hooves', in T. Ingold and J. L. Vergunst (eds.) *Ways of Walking: Ethnography and Practice on Foot*, Aldershot: Ashgate, 67–80.
Gosling, E. and K. J. H. Williams (2010). 'Connectedness to nature, place attachment and conservation behaviour: Testing connectedness theory among farmers', *Journal of Environmental Psychology* 30 (3): 298–304.
Gould, P. and R. White (1993). *Mental Maps*, London: Penguin.
Green, K. (2014). *Carp Fishing Manual: The Step by Step Guide to Becoming a Better Carp Angler*, Yeovil: J. Haynes Publications.
Gregory, D. (1994). *Geographical Imaginations*, Oxford: Blackwell.
Hale, C. (2006). 'Activist research v. cultural critique: indigenous land rights and the contradictions of politically engaged anthropology', *Cultural Anthropology* 21 (1): 96–120.
Hannerz, U. (2010). *Anthropology's World: Life in a Twenty-First Century Discipline*, London: Pluto Press.
Haraway, D. (2003). *A Companion Species Manifesto: Dogs, People and Significant Other-ness*, Chicago: Chicago University Press.
—— (2008). *When Species Meet*, Minneapolis: University of Minnesota Press.
Hardin, R. (2011). 'Collective contradictions of "corporate" environmental conservation', *Focaal* 60: 47–60.
Hebdige, D. (1990). 'Introduction – subjects in space', *New Formations* 11: vi-vii.
Heidegger, M. (2002). *Basic Writings*, London: Routledge.
Hicks, D. and M. Beaudry (2010). *The Oxford Handbook of Material Culture Studies*, Oxford: Oxford University Press.
Hilson, G. (2006). 'Improving environmental and ethical performance in the mining industry: Part 1: Environmental management and sustainable development', *Journal of Cleaner Production* 14 (3–4): 225–6.
Hines, J. M., H. R. Hungerford and A. N. Tomera (1987). 'Analysis and synthesis of research on responsible environmental behavior: a meta-analysis', *The Journal of Environmental Education* 18 (2): 1–8.
Hirsch, E. and M. O'Hanlon (eds.) (1995). *The Anthropology of Landscape*, Oxford: Oxford University Press.
Hodder, I. (2012). *Entangled: An Archaeology of the Relationship Between Humans and Things*, Chichester: Wiley-Blackwell.
Hoelle, J. (2012). 'Black hats and smooth hands: Elite status, environmentalism, and work among the ranchers of Acre, Brazil'. *Anthropology of Work Review* 33 (2): 60–72.
Homewood, K., P. Kristjanson and P. C. Trench (2009). 'Changing land use, livelihoods and wildlife conservation in Maasailand' in K. Homewood, P. Kristjanson and P. C. Trench (eds.) *Staying Maasai? Livelihoods, Conservation and Development in East African Rangelands*, New York: Springer, 1–42.
Hornborg, A. and M. Kurkiala (eds.) (1998). *Voices of the Land*, Lund: Lund University Press.

Horowitz, L. S. (2006). 'Mining and sustainable development', *Journal of Cleaner Production* 14 (3–4): 307–8.

—— (2010). '"Twenty years is yesterday": Science, multinational mining, and the political ecology of trust in New Caledonia', *Geoforum* 41 (4): 617–26.

Horton, D. and J. Parkin (2013). 'Conclusion: towards a revolution in cycling' in J. Parkin (ed.) *Cycling and Sustainability*, Bingley: Emerald Press, 303–25.

Horton, D., P. Cox and P. Rosen (2007). 'Introduction: cycling and society', in D. Horton, P. Cox and P. Rosen (eds.) *Cycling and Society*, Aldershot: Ashgate Publishing Company, 1–23.

Howes, D. (ed.) (1991). *The Varieties of Sensory Experience*, Toronto: University of Toronto Press.

—— (2005). *Empire of the Senses: The Sensual Culture Reader*, Oxford: Berg.

Hsiao, H.-C. and H.-M. Chen (2012). 'Keep walking: walking as detour from 'fitness' and the building of self-in-isolation and identity in contemporary consumer society', *Sport in Society* 15 (10): 1426–31.

Hunn, S. (2012). *Humans and Other Animals*, London: Pluto Press.

Ihde, D. (1990). *Technology and the Lifeworld*, Bloomington: Indiana University Press.

—— (2002). *Bodies in Technology*, Minneapolis: University of Minnesota Press.

Indigenous Peoples' and Local Community Conserved Areas and Territories (ICCA Consortium). http://iccaconsortium.wordpress.com

Ingold, T. (1993). 'Globes and spheres: The topology of environmentalism' in K. Milton (ed.) *Environmentalism: The view from anthropology*, London: Routledge, 31–42.

—— (2000). *The Perception of the Environment: Essays in Livelihood, Dwelling and Skill*, London: Routledge.

—— (2007). *Lines*, London: Routledge.

—— (2010). 'Footprints through the weather-world: walking, breathing knowing'. *Journal of the Royal Anthropological Institute* 16 (S1): 121–39.

—— (2011). *Being Alive: Essays on Movement, Knowledge and Description*, Abingdon: Routledge.

—— (2013). *Making: Anthropology, Archaeology, Art and Architecture*, Abingdon: Routledge.

Ingold, T. and J. L. Vergunst (eds.) (2008). *Ways of Walking: Ethnography and Practice on Foot*, Aldershot: Ashgate.

International Union for Conservation Nature (IUCN) http://www.iucn.org/about/union/commissions/ceesp/topics/governance/

Ittelson, W. H. (1973). 'Environment perception and contemporary perceptual theory', in W. H. Ittelson (ed.), *Environment and Cognition*, New York: Academic Press, 141–54.

Jackson, M. (1995). *At Home in the World*, Durham: Duke University Press.

Jackson, M. (ed.) (1987). *Anthropology at Home*, London: Tavistock.

——. (1996). *Things as They Are: New Directions in Phenomenological Anthropology*. Bloomington: Indiana University Press.

Jarowski, M. and T. Ingold (eds.) (2012). *Imagining Landscapes*, Farnham: Ashgate.

Johansson, M. and T. Hartig (2011). 'Psychological benefits of walking: moderation by company and outdoor environment', *Applied Psychology: Health and Well-being* 3 (3): 261–80.

Johnson, M. (2007). *The Meaning of the Body: Aesthetics of Human Understanding*, Chicago: University of Chicago Press.

Joppa, L. N., S. R. Loarie and S. L. Pimm (2009). 'On population growth near protected areas', *PLoS ONE* 4 (1): e4279 1–5. doi:10.1371/journal.pone.0004279.

Kaplan, R. (1973). 'Some psychological benefits of gardening', Environment and Behaviour 5: 145–52.

Kastner, J. and B. Wallis (1998). *Land and Environmental Art*, London and New York: Phaidon Press.

Kelly, K. and H. Francis (1994). *Navajo Sacred Places*, Bloomington: Indiana University Press.

Kirsch, S. (2002). 'Anthropology and advocacy: A case study of the campaign against the Ok Tedi Mine', *Critique of Anthropology* 22 (2): 175–200.

—— (2007). 'Indigenous movements and the risks of counterglobalization: Tracking the campaign against Papua New Guinea's Ok Tedi mine', *American Ethnologist* 34 (2): 303–21.

—— (2010a). 'Sustainability and the BP oil spill', *Dialectical Anthropology* 34 (3): 295–300.

—— (2010b). 'Sustainable mining', *Dialectical Anthropology* 34 (1): 87–93.

Kloss, F. R., T. Tuli, O. Haechl and R. Gassner (2006). 'Trauma injuries sustained by cyclists', *Trauma* 8 (2): 77–84.

Kottak, C. P. (1999). 'The new ecological anthropology', *American Anthropologist* 101 (1): 23–35.

Kremen, C., V. Razafimahatratra, R. P. Guillery, J. Rakotomalala, A. Weiss and J.-S. Ratsisompatrarivo (1999). 'Designing the Masoala National Park in Madagascar based on biological and socioeconomic data', *Conservation Biology* 13 (5): 1055–68.

Krueger, L. (2009). 'Protected areas and human displacement: Improving the interface between policy and practice', *Conservation and Society* 7 (1): 21–5.

Küchler, S. (2006). 'Reflections on art and agency: knot-sculpture between mathematics and art' in A. Schneider and C. Wright (eds.) *Contemporary Art and Anthropology*, Oxford and New York: Berg.

Küchler, S. and D. Miller (eds.) (2005). *Clothing as Material Culture*, Oxford: Berg.

Lake, S., J. Bullock and S. Hartley (2001). *Impacts of Livestock Grazing on Lowland Heathland in the UK*, Peterborough: English Nature.

Lakoff, G. and M. Johnson (1990). *Metaphors We Live By*, Chicago: Chicago University Press.

—— (1999). *Philosophy in the Flesh*, New York: Basic Books.

Lande, B. (2007). 'Breathing like a soldier: culture incarnate' in C. Shilling (ed.) *Embodying Sociology*, Oxford: Blackwell.

Landry, D. (2001). *The Invention of the Countryside: Hunting, Walking and Ecology in English Literature*, Basingstoke: Palgrave, 1671–831.

Lane, D. (2011). *Big Carp Legends*, Flitwick: Bountyhunter Publications.

Lang, N. (1991). *Nunga und Goonya*. München: Kunstraum München.

Lanting, H. (2014). *'Comparing and learning from each other for a better cycling future'*, Copenhagen: Networked Urban Mobilities Conference.

Latour, B. (1993). *We have Never Been Modern*, Cambridge, MA: Harvard University Press.

—— (2004). 'How to talk about the Body? The normative dimension of social science studies', *Body and Society* 10 (2–3): 205–29.

Laviolette, P. (2007). 'Hazardous sport?', *Anthropology Today* 23 (6): 1–2.

—— (2011a). *The Landscaping of Metaphor and Cultural Identity*, Frankfurt am Main: Peter Lang.

—— (2011b). *Extreme Landscapes of Leisure: Not a Hap-Hazardous Sport*, Farnham: Ashgate.

Lawrence, E. (1988). 'Horses in society' in A. Ronen (ed.) *Animals and People Sharing the World*, Hanover: University Press of New England, 95–114.

Leder, D. (1980). *The Absent Body*, Chicago: University of Chicago Press.

Lee, J. and T. Ingold. (2006). 'Fieldwork on foot: perceiving, routing, socializing' in S. Coleman and P. Collins (eds.) *Locating the Field: Space, Place and Context in Anthropology*, Oxford: Berg, 67–85.

Lemonnier, P. (2013). *Mundane Objects*, Walnut Creek, CA: Left Coast Press.

Leverington, F., K. L. Costa, J. Courrau, H. Pavese, C. Nolte, M. Marr, L. Coad, N. Burgess, B. Bomhard and M. Hockings (2010). *Management Effectiveness Evaluation in Protected Areas – A Global Study* (2nd edition), Brisbane: University of Queensland. Online. www.wdpa.org/me/PDF/global_study_2nd_edition.

Li, T. M. (2010). 'Indigeneity, capitalism, and the management of dispossession', *Current Anthropology* 51 (3): 385–414.

Little, P. E. (1999). "Environments and environmentalisms in anthropological research: facing a new millenniuim', *Annual Review of Anthropology* 28, 253–84.

Lloyd, A. (n.d.). 'The contemporary art of walking'. Online. www.fermynwoods.co.uk/archive/workshopstalksevents/alison-lloyd-the-contemporary-art-of-walking/

Long, R. (1980). *Five, Six, Pick Up Sticks Seven, Eight, Lay Them Straight*. London: Anthony d'Offay Gallery.

—— (1994). *No Where. Interview with Colin Kirkpatrick*. 8 July 1994. Piers Art Centre, Orkney. Online. www.speronewestwater.com/cgi-bin/iowa/articles/record.html?record=293. Cited by Karen O'Rourke (2013). *Artists as Cartographers*. Cambridge, MA: MIT Press, 61.

Low, S. M. and S. E. Merry. (2010). 'Engaged anthropology: diversity and dilemmas', *Current Anthropology* 51(S2): S203–25.

Lowndes, S. (2014). *All Art is Political. Writings on Performative Art.* Edinburgh: Luath Press Limited.

Lund, K. (2012). 'Landscapes and narratives: compositions and the walking body'. *Landscape Research* 37 (2): 225–37.

MacNaughton, P. and J. Urry (1998). *Contested Natures,* London: Sage.

Marioni, T. (1988). 'Chris Burden: A sculptor's sensibility, the early years' in A Ayres and P. Schimmel (eds.) *Chris Burden, a Twenty Year Survey,* Newport: Newport Harbor Art Museum. Cited by G. Marvin and S. McHugh (eds.) (2014). *Routledge Handbook of Human-Animal Studies,* London: Routledge.

Marratto, S. (2012). *The Intercorporeal Self,* Albany: SUNY Press.

Marvin, G. and S. McHugh (2014). *Routledge Handbook of Human-Animal Studies,* London: Routledge/Taylor and Francis.

Matilsky, B. C. (1992). *Fragile Ecologies: Contemporary Artists' Interpretations and Solutions,* New York: Queens Museum of Art, Rizzoli International.

Matless, D. (1998). *Landscape and Englishness,* London: Reaktion.

Maurstad, A., D. Davis and S. Cowles (2013). 'Co-being and intra-action in horse-human relationships: a multi-species ethnography of be(com)ing human and be(com)ing horse', *Social Anthropology* 21(3): 322–35.

Mauss, M. (1979) [1935]. 'Techniques of the body' in M. Mauss *Sociology and Psychology: Essays by Marcel Mauss (trans. B. Brewster),* London: Routledge Kegan Paul, 95–123.

Maxwell, N., O. Gibson and R. Twomey (2012). *The Health Benefits of Horse Riding in the UK,* Kenilworth: British Horse Society.

Mbile, P., M. Vabi, M. Meboka, D. Okon, J. Arrey-Mbo, F. Nkongho and E. Ebong (2005). 'Linking management and livelihood in environmental conservation: case of the Korup National Park Cameroon', *Journal of Environmental Management* 76 (1): 1–13.

McAndrew, J. P. and Il, O. (2004). *Upholding Indigenous Access to Natural Resources in Northeast Cambodia,* Phnom Penh: Asian Development Bank.

McKenney, B., Y. Chea, P. Tola, and T. Evans (2004). *Focusing on Cambodia's High Value Forests: Livelihoods and Management.* Phnom Penh: Cambodia Development Resource Institute and Wildlife Conservation Society, 71–2.

McNab, R. B. and V. H. Ramos (2007). 'The Maya biosphere reserve and human displacement: Social patterns and management paradigms under pressure', in K. H. Redford and E. Fearn (eds.) *Protected Areas and Human Displacement: A Conservation Perspective,* Bronx, NY: WCS Institute, 20–27. Online. www2.fiu.edu/~brayd/wcswp29.pdf

Meinig, D. (ed.) (1979). 'The beholding eye' in D. Meinig *The Interpretation of Ordinary Landscapes,* Oxford: Oxford University Press, 33–48.

Merleau-Ponty, M. (1962). *The Phenomenology of Perception,* London: Routledge.

—— (1968). *The Visible and the Invisible,* Evanston, Northwestern University Press.

Metken, G. (1991). in Lang, N. (1991). *Nunga und Goonya.* München: Kunstraum München, 34–85.

Miller, D. (ed.) (1998). *Material Cultures: Why Some Things Matter*, London: UCL Press.
—— (2005). *Materiality*, Durham: Duke University Press.
—— (2010). *Stuff*, Cambridge: Polity Press.
Milton, K. (1991). 'Interpreting environmental policy: a social scientific approach', *Journal of Law and Society*, 18 (1): 4–17.
—— (2002). *Loving Nature*, London: Routledge.
Mines and Communities (MAC) http:/minesandcommunities.org/
Morgan, D. (1994). 'Theatre of war: combat, the military and masculinities' in H. Brod and M. Kaufman (eds.) *Theorising Masculinities*, London: Sage, 165–83
Morita, E., S. Fukuda, J. Nagano, N. Hamajima, H. Yamamoto, Y. Iwai, T. Nakashima, H. Ohira and T. Shirakawa (2007). 'Psychological effects of forest environments on healthy adults: "Shinrin-yoku" (forest-air bathing, walking) as a possible method of stress reduction', *Public Health* 121 (1): 54–63.
Morphy, H. (1993). 'Colonialism, history and the construction of place: the politics of landscape in northern Australia' in B. Bender (ed.) *Landscape: Politics and Perspectives*, Oxford: Berg, 205–44.
Morris, B. (2015). *The Walk Exchange*. Online. www.walkexchange.org
Morrison-Bell, C. (2013). Foreword in C. Morrison-Bell, M. Collier, with assistance from J. Ross (eds.) *Walk On: From Richard Long to Janet Cardiff – 40 Years of Art Walking*, Art Editions North, Manchester: Cornerhouse Publications, 1–3 and http://walk.uk.net/portfolio/walk-on/
Moskos, C., J. Williams and D. Segal (2000). *The Postmodern Military: Armed Forces after the Cold War*, Oxford: Oxford University Press.
Munn, N. (1973). 'The spatial presentation of cosmic order in Walbiri iconography' in A. Forge (ed.) *Primitive Art and Society*, Oxford: Oxford University Press.
Naji, M. (2009). 'Gender and materiality in-the-making: the manufacture of Sirwan femininities through weaving in southern Morocco', *Journal of Material Culture* 14 (1): 47–73.
Nelson, R. (1983.) *Make Prayers to the Raven*, Chicago: Chicago University Press.
Newsinger, J. (1997). *Dangerous Men: the SAS and Popular Culture*, London: Pluto Press.
Nielsen, M. R. and T. Treue (2012). 'Hunting for the benefits of joint forest management in the Eastern Afromontane biodiversity hotspot: Effects on bushmeat hunters and wildlife in the Udzungwa Mountains', *World Development* 40 (6): 1224–39.
Nielsen, N. K. (2003). 'New Year in Nämpnäs: on nationalism and sensuous holidays in Finland', *Tourist Studies* 3 (1): 83–98.
Nunan, F. (2015). *Understanding Poverty and the Environment: Analytical Frameworks and Approaches*, Oxon: Routledge.
Obi, C. I. (1997). 'Globalisation and local resistance: The case of the Ogoni versus Shell'. *New Political Economy* 2 (1): 137–48.
Okely, J. and H. Callaway (eds.) (1992). *Anthropology and Autobiography*, London: Routledge.

Olsen, B. (2010). *In Defense of Things*, Lanham: AltaMira Press.
Office for National Statistics (2011). www.neighbourhood.statistics.gov.uk
Ozinga, S. (2003). *Parks with People*. World Rainforest Movement/FERN. Online. http://www.fern.org/sites/fern.org/files/pubs/ngostats/parks.htm
Patinkin, J. (2013). 'Tanzania's Maasai battle game hunters for grazing land'. BBC News, 18 April. www.bbc.co.uk/news/world-africa-22155538
Pearce, F. (2005). 'Big game losers'. *New Scientist*, 16 April: 21.
Pence, G. Q. K, M. A. Botha and J. K. Turpie (2003). 'Evaluating combinations of on-and off-reserve conservation strategies for the Agulhas Plain, South Africa: a financial perspective', *Biological Conservation* 112 (1–2): 253–73.
Pendleton, A., S. McClenaghan, C. Melamed, I. Bunn and D. Graymore (eds.) (2004). *Behind the Mask: The Real Face of Corporate Social Responsibility*, London: Christian Aid.
Phelan, P. (1993). *Unmarked: The Politics of Performance*, London: Routledge.
Pink, S. (2009). *Doing Sensory Ethnography*, London: Sage.
Pinney, C. (2001). 'Piercing the skin of the idol' in C. Pinney and N. Thomas (eds.) *Beyond Aesthetics: Art and the Technologies of Enchantment*, Oxford, New York: Berg, 157–79.
—— (2002). 'Creole Europe: The reflection of a reflection', *Journal of New Zealand Literature* 20: 125–61.
Pinney, C. and N. Thomas (eds.) (2001). *Beyond Aesthetics. Art and the Technologies of Enchantment*, Oxford: Berg.
PLaCE (2012). *Between: Ineffable Intervals*, Bristol: The Little Printing Company.
Pope, S. (2012). 'The memorial walks', in D. Evans (ed.) *The Art of Walking: a Field Guide*, London: Black Dog Publishing, 57–61.
Probert, G. B. (2004). 'An Investigation into whether Mountain Biking Facilitates Escapism and Mental Freedom'. BA thesis, University of Cumbria. Published on the International Mountain Bicycling Association website. *imba.org.uk/wp-content/uploads/MTB_Escapism_Research.pdf*
Public Eye (2011). http://www.publiceye.ch
Rapport, N. (1993). *Diverse World Views in an English village*, Edinburgh: Edinburgh University Press.
—— (ed.) (2002). *British Subjects: An Anthropology of Britain*, London: Bloomsbury.
—— (2008) 'Walking Auschwitz, walking without arriving', *Journeys* 9 (2): 32–54.
—— (2010) 'The ethics of participant observation: personal reflections on fieldwork in England' in P. Collins and A. Gallinat (eds.) *The Ethnographic Self as Resource*, Oxford: Berghahn.
Rapport, N. and Dawson, A. (eds.) (1998). *Migrants of Identity: Perceptions of Home in a World of Movement*, Oxford: Berg.
Ravenscroft, N. O. M. and M. S. Warren (compilers) (1996). *Species Action Plan. The Silver-Studded Blue (Plebejus argus)*, Wareham: Butterfly Conservation.
Reid, H., D. Fig, H. Magome and N. Leader-Williams (2004). 'Co-management of contractual national parks in South Africa: Lessons from Australia', *Conservation and Society* 2 (2): 377–409.

Rival, L. (2011). 'Anthropological encounters with economic development and biodiversity conservation', Working Paper Number 186, Series: QEHWPS:QEHWPS186, Oxford Department of International Development. Online. www.qeh.ox.ac.uk/publications

Robertson, R., A. Robertson, R. Jepson and M. Maxwell (2012). 'Walking for depression or depressive symptoms: a systematic review and meta analysis', *Mental Health and Physical Activity* 5: 66–75.

Rosen, P. (2002). *Framing Production: Technology, Culture, and Change in the British Bicycle Industry*, Cambridge, MA: MIT Press.

Rubinstein, D. and C. Speakman (1969). *Leisure, Transport and the Countryside*, London: Fabian Society.

Ruggie, J. (2008). *Protect, Respect and Remedy: a Framework for Business and Human Rights, UN Human Rights Council*. Online. www.reports-and-materials.org/sites/default/files/reports-and-materials/Ruggie-report-7-Apr-2008.pdf

Sahlins, M. (1957). 'Land use and the extended family of Moala, Fiji', *American Anthropologist* 59 (3): 449–62. Cited by Scoones, I. (1999). 'New ecology and the social sciences: what prospects for a fruitful engagement?', *Annual Review of Anthropology* 28: 479–507.

Samimian-Darash, L. (2012). 'Rebuilding the body through violence and control', *Ethnography* 14 (1): 46–63.

Sands, R. R. (2010). 'Anthropology revisits sport through human movement', in Sands, R. R. and L. R Sands (eds.) *The Anthropology of Sport and Human Movement: A Biocultural Perspective*, Lanham; Plymouth: Lexington Books, 5–37.

Sasson-Levy, O. (2007). 'Research on gender and the military in Israel: From a gendered organization to inequality regimes', *Israel Studies Review*, 26 (2): 73–98.

Schacter, R. (2014). *Ornament and Order: Graffiti, Street Art and the Parergon*. Farnham: Ashgate Publishing Limited.

Scheper-Hughes, N. and M. Lock (1987). 'The mindful body: a prolegomenon to future works in medical anthropology', *Medical Anthropology Quarterly* 1 (1): 6–41.

Schneider, A. and C. Wright (eds.) (2006). *Contemporary Art and Anthropology*, Oxford; New York: Berg.

—— (2010). *Between Art and Anthropology: Contemporary Ethnographic Practice*, London: Bloomsbury.

—— (2013). *Anthropology and Art Practice*, London: Bloomsbury.

Scoones, I. (1999). 'New ecology and the social sciences', *Annual Review of Anthropology* 28: 479–507.

Scott, M. and S. Johnson (2006). *The Battle for Roineabhal: Reflections on the successful campaign to prevent a superquarry at Lingerabay, Isle of Harris and lessons for the Scottish planning system*, Perth: Scottish Environmental LINK.

Sheets-Johnstone, M. (1999). *The Primacy of Movement*, Philadelphia: John Benjamins.

Short, C. and Hayes, E. (2005). *A Common Purpose: A Guide to Agreeing Management on Common Land*, Gloucester: Gloucester University Press.

Soja, E. (1989). *Postmodern Geographies: The Insertion of Space in Critical Social Theory*, London: Verso.

Solnit, R. (2001). *Wanderlust. A History of Walking*, London: Verso.

Sookermany, A. (2011). 'The embodied soldier: towards a new epistemological foundation of soldiering skills in the (post) modernized Norwegian armed forces', *Armed Forces and Society* 37 (3): 469–93.

SPARC (2003). *SPARC Facts: Results of the New Zealand Sport and Physical Activity Surveys, 1997–2001*. Online. www.srknowledge.org.nz/research-completed/sparc-facts-results-of-the-new-zealand-sport-and-physical-activity-surveys-1997-2001//

Speed, S. (2007). *Rights in Rebellion: Indigenous Struggle and Human Rights in Chiapas*, Stanford: Stanford University Press.

Spinney, J. (2006). 'A place of sense: a kinaesthetic ethnography of cyclists on Mont Ventoux', *Environment and Planning D: Society and Space* 24 (5): 709–32.

Stewart, P. and A. Strathern (eds.) (2003). *Landscape, Memory and History: Anthropological Perspectives*, London: Pluto Press.

Stokes, R. (1999). *The Book of Woodbury: The Twentieth Century Revisited*, Halsgrove Press: Tiverton.

Stoller, P. (1989). *The Taste of Ethnographic Things*, Philadelphia: University of Philadelphia Press.

Strang, V. (1997). *Uncommon Ground: Cultural Landscapes and Environmental Values*, Oxford: Berg, 199–215.

Strathern, M., A. Richards and F. Oxford (1981). *Kinship at the Core: An Anthropology of Elmdon, a Village in North-west Essex in the Nineteen-Sixties*, Cambridge: Cambridge University Press.

Strathern, M. (ed.) (2000). *Audit Cultures: Anthropological Studies in Accountability, Ethics and the Academy*, London: Routledge.

Survival International (2013). 'Brazilian Indians forced to leave mega-dam site'. 13 May. Online. www.survivalinternational.org/news/9220

Tam, K.-P. (2013). 'Concepts and measures related to connection to nature: Similarities and difference', *Journal of Environmental Psychology*. 34 (1): 64–78.

Tanner, A. (1979). *Bringing Home Animals*, Newfoundland: Institute of Social and Economic Research.

The Saint Consulting Group http://tscg.biz/saintblog/2009/03/quarries-hit-rock-bottom-as-most-opposed-property-sector-in-britain-saint-index.html

Tilley, C. (1994). *A Phenomenology of Landscape: Places, Paths and Monuments*, Oxford: Berg.

—— (1999). *Metaphor and Material Culture*, Oxford: Blackwell.

—— (2002). 'Metaphor, materiality and interpretation' in Buchli, V. (ed.) *The Material Culture Reader*. Oxford: Berg, 23–6.

—— (2004). *The Materiality of Stone*, Oxford: Berg.

—— (ed.) (2006). Landscape, Heritage and Identity, Special Double Issue of the *Journal of Material Culture* (Vol 11:1/2).

—— (2006a). 'Objectification' in C. Tilley, W. Keane, S. Küchler, M. Rowlands and P. Spyer (eds.) (2006). *The Handbook of Material Culture*. London: Sage, 60–73.

—— (2008). *Body and Image*, California, CA: Left Coast Press.

—— (2008a). 'From the English cottage garden to the Swedish allotment: banal nationalism and the concept of the garden', *Home Cultures* 5 (2): 219–49.

—— (2009). 'What gardens mean' in P. Vannini (ed.) *Material Culture and Technology in Everyday Life*, New York: Peter Lang, 171–92.

—— (2010). *Interpreting Landscapes*, Walnut Creek, CA: Left Coast Press.

—— (2012). 'Walking the past in the present' in A. Arnason, N. Ellison, J. Vergunst and A. Whitehouse (eds.) *Landscapes Beyond Land: Routes, Aesthetics, Narratives*, Oxford: Berghahn Books, 15–32.

—— (2017). *Landscape in the Longue Durée: A History and Theory of Pebbles in a Pebbled Heathland Landscape*, London: UCL Press.

Tilley, C., W. Keane, S. Küchler, M. Rowlands and P. Spyer (eds.) (2006). *The Handbook of Material Culture*, London: Sage.

Timura, C. T. (2001). '"Environmental conflict" and the social life of environmental security discourse', *Anthropological Quarterly* 74 (3): 104–13.

Torkar, G. and S. McGregor (2012). 'Reframing the conception of nature conservation management by transdisciplinary methodology: From stakeholders to stakesharers', *Journal for Nature Conservation* 20 (2): 65–71.

Tuan, Y.-F. (1974). *Topophilia*, New York: Columbia University Press.

—— (1977). *Space and Place*, Minneapolis: University of Minnesota Press.

Tuck-Po, L. (2008). 'Before a step too far: walking with Batek hunter-gatherers in the forests of Pahang, Malaysia', in T. Ingold and J. Lee Vergunst (eds.) *Ways of Walking: Ethnography and Practice on Foot*, Aldershot: Ashgate, 67–80.

Turner, T. (1995). 'An indigenous peoples struggle for socially equitable and ecologically sustainable production: the Kayapo revolt against extractivism', *Journal of Latin American Anthropology* 1 (1): 98–121.

Underhill-Day, J. (2009). The Pebblebed Heaths: an Options Appraisal. Online. www.pebblebedheaths.org.uk/_assets/full_consultation_report_without_appendices_200509.pdf

Van Den Broek, G. J. (1984). 'The sign of the fly: a semiotic approach to fly-fishing in Britain', *American Journal of Semiotics* 3 (1): 71–79.

Vancouver, C. (1969) [1808]. *General View of the Agriculture of the County of Devon*, Newton Abbot: David and Charles.

Vergunst, J. (2007). 'Moving in nature: Walking and completedness in nature'. Paper presented at 'Performing Nature at the World's Ends', University of Oslo, 29–31 August.

Vilaça, A. (2009). 'Bodies in perspective: a critique of the embodiment paradigm from the point of view of Amazonian ethnography' in H. Lambert and M. McDonald (eds.) *Social Bodies*, Oxford: Berghahn: 129–47.

Vivanco, L. (2013). *Reconsidering the Bicycle: An Anthropological Perspective on a New (Old) Thing*, London: Routledge.

Warnier, J.-P. (2001). A praxeological approach to subjectification in a material world', *Journal of Material Culture* 6 (1): 5–24.

Watts, M. (1993). 'Development I: power, knowledge and discursive practice', *Progress in Human Geography* 17 (2): 257–72. Cited by West, P., J. Igoe and D. Brockington (2006). 'Parks and peoples: the social impact of protected areas', *Annual Review of Anthropology* 35: 251–77.

Ween, G. and S. Abram (2012). 'The Norwegian Trekking Association: trekking as constituting the nation', *Landscape Research* 37 (2): 155–71.

Weiner, J. (1991). *The Empty Place: Poetry, Space and Being among the Foi of Papua New Guinea*, Bloomington: Indiana University Press.

Weiss, M. (2002). *The Chosen Body: The Politics of the Body in Israeli Society*, Stanford: Stanford University Press.

Welker, M. A. (2009). 'Corporate security begins in the community': mining, the corporate social responsibility industry, and environmental advocacy in Indonesia', *Cultural Anthropology* 24 (1): 42–179.

West, P., J. Igoe and D. Brockington (2006). 'Parks and peoples: the social impact of protected areas'. *Annual Review of Anthropology* 35: 251–77.

White, G. (1977) [1788–9]. *The Natural History of Selborne*, Harmondsworth: Penguin.

Whitmore, A. (2006). 'The emperor's new clothes: Sustainable mining?', *Journal of Cleaner Production* 14 (3–4): 309–14.

Widlock, T. (2008). 'The dilemmas of walking: a comparative view', in T. Ingold and J. Lee Vergunst (eds.) *Ways of Walking: Ethnography and Practice on Foot*, Aldershot: Ashgate, 51–66.

Wilson, M. (2004). *A Woodbury Triumph!*, Paulton: MLD Litho and Digital Print.

Winslow, D. (1999). 'Rites of passage and group bonding in the Canadian airborne', *Armed Forces and Society* 25 (3): 429–57.

Wittemyer, G., P. Elsen, W. T. Bean, A. C. O. Burton, and J. S. Brashares (2008). 'Accelerated human population growth at protected area edges', *Science* 321: 123–26.

Wittgenstein, L. (1981) [1921]. *Tractatus Logico-Philosophicus*, Abingdon: Routledge.

Wohlwill, J. F. (1976). 'Environmental aesthetics: The environment as a source of affect', in I. Altaian and J. F. Wohlwill (eds.), *Human Behavior and Environment* (Vol. 1). New York: Plenum Press, 37–86.

Woodward, R. (1998). '"It's a man's life!": Soldiers, masculinity and the countryside', *Gender, Place and Culture* 5 (3): 277–30.

Wordsworth, W. (1888). 'The prelude. Book thirteenth. Imagination and taste, how impaired and restored' in *The Complete Poetical Works*. London: Macmillan and Co. Online. www.bartleby.com/145/

Wright, C. (2013). Discussion of AI's bid for Straitgate Farm. [email] (Personal communication 18 July 2013).

Yeung, A. B. (2004). 'The octagon model of volunteer motivation: Results of a phenomenological analysis'. *Voluntas: International Journal of Voluntary and Nonprofit Organizations* 15 (1): 21–46.

Zandrino, F., F. Musante, N. Mariani and L. E. Derchi (2004). 'Partial unilateral intracavernosal hematoma in a long-distance mountain biker: a case report', *Acta Radiologica* 45 (5): 580–3.

Zavestoski, S. (2003). 'Constructing and maintaining ecological identities: the strategies of deep ecologists', in S. Clayton and S. Opotow (eds.), *Identity and the Natural Environment: The Psychological Significance of Nature*. Cambridge, MA: MIT Press, 297–315.

Zeiske, C. (2014). 'A train of thought' in M. J. Jacob and C. Zeiske (eds.) *Fernweh: A Travelling Curators' Project*. Berlin: Jovis, 15–20.

Websites

Axe Valley Pedallers www.axevalleypedallers.org.uk/

Bicycle Helmet Research Foundation. www.cyclehelmets.org

Clinton Devon Estates www.clintondevon.com

Cycle Touring Club Exeter www.ctcdevon.co.uk

East Devon Area of Outstanding Natural Beauty www.eastdevonaonb.org.uk

EveryTrail: One cyclist has videoed himself and his friends riding mountain bikes on some of the single tracks on Woodbury Common; this may be seen on the following site. It gives a clear indication of the pleasure to be had in riding on the heathland: www.everytrail.com/view_trip.php?trip_id=1101382

Exeter Mountain Bike Club www.embc.uk.net

Exeter Outdoor Group (n.d.). www.eog.org.uk/

Fairlynch Museum www.devonmuseums.net/Fairlynch-Museum/Devon-Museums/

Indigenous Peoples' and Local Community Conserved Areas and Territories (ICCA Consortium). iccaconsortium.wordpress.com

International Union for Conservation Nature (IUCN) www.iucn.org/about/union/commissions/ceesp/topics/governance/

Landance: This non-profit organization runs workshops in contemporary dance, music, visual art and film, leading to performances in the landscape in the South West. www.landance.org.uk

Mines and Communities (MAC) http://minesandcommunities.org/

Moving Naturally: '"Moving Naturally" is a journey to access a natural ease of movement innate in babies, children and people. It is present in the evolution of human movement from life in water to life on land. Time spent in land and seascapes with its diverse wildlife provides an opportunity to access a more natural way of being and moving. It raises the question "How do I want to move?"' www.movingnaturally.co.uk

Open Spaces Society www.oss.org.uk/

Otter Valley Association: "The OVA is a Civic society founded to interest residents and visitors in the history, geography, natural history, architecture and future planning of this area of Devon." www.ova.org.uk
Ramblers' Association www.ramblers.org.uk
Royal Marines www.royalmarines.mod.uk/
Sid Valley Cycling Club www.svcc.org.uk
Walking Artists Network http://www.walkingartistsnetwork.org/

Index

Page numbers in italics are figures; with 't' are tables.

acrylic artist (Caroline) 245
aesthetics 3, 241; anti-aestheticism 241
agency 3, 8–9
Aggregate Industries 25, 152–3, 157–60, 161–54
Aggregate Levy Sustainable Fund 25, 157
analysis of research 297–8
Anderson, B. 19
anglers *see* fishermen
anthropology 20; of Britain 18–19; and cycling 178–9; of sport 176–7; and walking 215–16
Anthropology and Art Practice (Schneider and Wright, eds) 240
anti-aestheticism 241
archaeology/archaeologists 17–23, 294; conflict with heathland management objectives 59–64; perspective on the heathland 46–7; in the PHCT report 43; and swaling 65–6; and visitor survey 168–70; *see also* Trenchard, Priscilla; Woodbury Castle
art/artists 17–23, 234–5, *299*; and anthropology 240–2; artists 242–7, 261; heathland 241; nature and culture 235–6; and walking 236–9; *see also* acrylic artist (Caroline); Hearn, Barbara; Trenchard, Priscilla
Aylesbeare and Harpford Commons 25, 28, *34*, *51*, *52–3*; conflict with archaeology 59–64; fencing 72; grazing 68; and management of people on the heathland 79, 80–2; place names 36–7, *38–9*, *40–1*, 42; and the RM 89, 119; and top soil scraping 58, *59–65*; visitors 167–9

baby crawl 92
baits for carp fishing 268–9
Basso, K. 36
beach artist (Barbara Hearn) 248–51
Besnier, N. 177
Between Art and Anthropology (Schneider and Wright, eds) 240
bikes 179–81

birds 43–4, 46, 74, 130; Dartford warbler (*Sylvia undata*) 26, 50, 74, 118, 130, 140, 171–5; nightjars (*Caprimulgus europaeus*) 26, 139–42, 171–5; and the RM 118, 121; visitor awareness of 171
Black Hill quarry *see* quarrying
Borden, I. 179
Bourdieu, P. 178, 297
Brechin, S. R. 129–30
Brosius, J. P. 161
Brown, T. 152
Brownell, S. 177
burning *see* fire; swaling
butterflies 63, 133, 137–8, 171–5, 290
Buxton, Richard 213

Cameron-Daum, Kate 61, 165, 214
Cándida-Smith, R. 234
car parks 79–80
carp 264–5, 267, 269–71
Carrier, J. 12
Carter, George 65; memorial *292–9*
Carter, P. 45
cattle 33–5, 137, 139
CDE *see* Clinton Devon Estates (CDE)
Chadwick, Gregg 238
cherry trees 35–7
clearance, land 127
climate change 48
Clinton Devon Estates (CDE) 25–30; and commoners' rights to graze 74; and fencing 71; and the model flyers club 275; and quarrying 153, 155; and the RM 89
Clinton family 28
Clothier, P. 238
coarse fishing 262
Colaton Raleigh Common *49–51*, *52–3*, *65–7*, 167–9; archaeology 294; visitors 167–9
collaboration 241
Collier, Mike 239
common land 74–5
'Common Purpose' strategy 71

concealment of the body 91–2
conflict/contestation 3, 9–10, 56–8, 82–3, 126, 295–8
connectivity with nature (CWN) 128
conservation, and RM 116–21
conservation grazing 71, 72, 75
consultation report, heathland fencing 68–73
Contemporary Art and Anthropology (Schneider and Wright, eds) 240
contestation/conflict *see* conflict/contestation
Corporate Environmental and Social Responsibility (CESR) practice 152–3
Cosgrove, D. 4
Countryside and Rights of Way Act (2000) 68–9, 75
Cox, P. 175–6
Csikszentmihaly, M. 115
culture 3, 19, 235–6
CWN *see* connectivity with nature (CWN)
cyclists/cycling 17–23, 169–71, 175–9, 210–12; and anthropology 178–9; at night 190–2; awareness of wildlife 171; bikes/apparel 179–81; comparison to walking 192–3; effect on the landscape 199–200; and the landscape 193–8; relationship with other users of the landscape 198–200; riding groups 183–9; routes 187–8, *189–202*; techniques of riding 181–3; and walkers 226

Dalditch World War II camp 39, 50, 63, 87
damselflies (*Coenagrion mercuriale*) 138–9, 171–5
dancer (Michelle) 245–7
danger in nature 289
Daniels, S. 4
Dartford warbler (*Sylvia undata*) 26, 50, 74, 118, 130, 140, 171–5; and Carter memorial *292–9*
Dawson, A. 232
De Nardi, S. 13
Dean, Margaret 242–52
Deleuze, G. 216
description of research 297–8
Deveron Arts (Scotland) 236–7
Devon Wildlife Trust (DWT) 25
Dillon, B. 236, 237
disciplinary time 93
discipline 85, 102
dogs/dog walkers 166–7, 169–71, 226–8; and coarse fishing 264; and the heathland 75–7, 117–18, 171; and the model flyers club 275
Dufrenne, M. 234–5
Dutcher, D. D. 128
dwelling perspective 216
Dyck, N. 177

East Budleigh Common 42, 56, 121, 167–9, 264
East Devon Area of Outstanding Natural Beauty (AONB) 26
East Devon Radio Control Club (EDRCC) 273, 274–6
Eliot, T. S. 234
embodiment 6–9, 290–1; and art 261; and horse riding 209–10; and model aircraft flyers 285; and the RM 121–4; and visitors 174; *see also* cyclists/cycling; training programme of the Royal Marines; walkers/walking
emotion 10–13, 172, 214, 288, 289–90; and art 238, 251, 258, 291; and bikers' experiences 195; and fishermen 271; and place 291–4; and RM training 100; and the volunteers 150–1; and walking 217
Empson, R. 241
endurance course 89, *106*, 107–16
environmentalists/environmentalism 17–23, 125–6; politics of 129–30; *see also* volunteers

fading, from conscious experience 115
Farley, Barbara 243–4; 'Watching for Nightjars' (poem) 243–5
favourable/unfavourable heathland 44–6, 49–50, 58, 68
fencing 137; consultation report 68–73; Woodbury Common training ground 87
fire; and art 256–8; and model aircraft flyers 284; wild 254–5
firebreaks *52–3*, 54, 78
fishing/fishermen 17–23, 262–72
flow, state of consciousness 115
forestry 28, 29, 30, 127
Foucault, M. 85, 93, 124, 224
Fowler, C. 177
Francis, H. 36
Frynas, Jedrzej 127
Fulton, H. 239

Gaffin, D. 36
Geertz, C. 125, 176
Geismar, H. 241
Gell, A. 235, 241
geology, in the PHCT report 43
ghost walk 92–3
Goldsworthy, Andy 238–9
golf courses 29, 153, 217–18, 223
Gooch, Pernille 215
gorse 131, 169–71, 203, 227, 253; in art 256, *257–8*; management of 46, 50, 62, *79–45*; and model aircraft flying 284; and Royal Marine training 79, 104–5, 122
Gosling, E. 128
Gotham, Pete 31, 33, 50, 58, 66
grazing 33–5, 66–8, 71; *see also* fencing

grenade range 87, *88–96*
guides for visitors 78–9

'harbour areas' 91, *101–6*
Hardin, Rebecca 128
Harpford Common *see* Aylesbeare and Harpford Commons
Hearn, Barbara 248–51, *252–7*
Heath Week celebration 81–2, 139, 143–4
heathlands 130
Hebdige, D. 232
Heidegger, M. 20
Higher Level Stewardship Scheme (HLS) 26, 30, 46–7, 60
Hilson, G. 152
historic environment *see* archaeology
HLS *see* Higher Level Stewardship Scheme (HLS)
Horner, Bill 61, 62
Horowitz, Leah 127–8
horse riding 17–23, 175, *183*, 201–9, 210–12; effect on landscape 199; embodiment 209–10; and other heathland users 209
Horton, D. 177–8
hunting, trail 219–20

identity 5, 290; ecological 12–13
indigenous and community conserved areas (ICCAS) 129
indigenous traditional societies 11–12
informants 16–23
Ingold, Tim 5, 125, 216; *Ways of Walking* 215
insects 130–1; butterflies 63, 133, 137–8, 171; damselflies (*Coenagrion mercuriale*) 138–9, 171–5
International Union for Conservation of Nature (IUCN) 129
Isle of Harris (Western Isles, Scotland), Lingeraby quarry 161

Johnson, M. 8, 10–11
Joint Nature and Conservation Council (JNCC) 46

Kelly, K. 36
kit, Royal Marine 100–1
kitten crawl 92
Krueger, L. 129, 150
Küchler, S. 235

Lafarge Redland Aggregates 161
Lakoff, G. 8
land ownership 11
Lang, N., *Nunga und Goonya* 240–1
Latour, B. 6, 7
Laviolette, P. 2, 36, 177
Leder, D. 115
leopard crawl 92

Lingeraby quarry, Isle of Harris (Western Isles, Scotland) 161
litter 80; and the RM 117; *see also* dogs
Little, Paul 125
livestock 54
Lloyd, A. 236
Long, R. 239
Lowndes, S. 236

management 31–6, 48–56, 128–9; of people on the heathland 73–5; public access 77–82; techniques 69, *see also* grazing; scraping, top soil; swaling
maps 293; by cyclists 188, *189–202*, 211; by fishermen 265–7; by horse riders 204–7, 211; by model aircraft flyers 282, *283*; by Royal Marines 293; by volunteers 145, *146–7*; by walkers 229, *230–1*; Natural England *57*
Marx, Karl 287, 288, 297
material culture studies 5–6
materiality 4–6, 287–8, 296–7
Matilsky, Barbara C. 235–6
Matless, D. 4
Mauss, M. 84–5, 124, 215
meaning 2, 178, 241, 288, 289, 295–6, 297–8; and contestation 295–8; and emotion 10–11, 291–2
mediation 3
memory 18, 292–3, 294–5; and art 237, 261, *299*; and emotion 13; and horse riders 212; and model aircraft flyers 285–6; and the RM 121–4; and walkers 214, 232
Merleau-Ponty, M. 6–7, 239
Metken, G. 241
Milton, Kay 11–12, 126, 150, 161
model aircraft flyers 17–23, 273–4, *278*, *281–3*, 285–78; East Devon Radio Control Club (EDRCC) 273, 274–6; and the environment 282–5; flying 279–82; planes 276–9; and walkers 226
monkey run 92
Morrison-Bell, C. 239
mounds, landscaping 55–6
mountain biking 178
movement through the landscape 3, 232; RM 92–4

names; of carp 269–70; *see also* place names
nationalism 19
Natural England (NE) 26, 42–6, 48, 49–50; on grazing 68; map *57*; and top soil scraping 61
Natural History of Selborne (White) 64
nature/natural 3, 223, 288–90; and culture 235–6; politics of 151, 289, 290
NE *see* Natural England (NE)

nightjar (*Caprimulgus europaeus*) 26, 139–42, 171–5
nutrification 76

officials 17–23
ordinariness 296
organic farming 28
ownership; sense of 144, 148, 150, 225, 283; *see also* Clinton Devon Estates (CDE)

panoramic sketching 94–5
participant observation 5, 18
participation 241
pebble structures, people with 17–23
Pebblebed heathland (Devon) 14–16
Pebblebed Heathlands Conservation Trust (PHCT) 25, 29–30, 33–6; landscaping after quarrying 155; place names 37–9, 42; and public access 81–2; report (2009) 43–4; Underhill-Day report 69–70; walking guide 78
people management on the heathland 73–5
perception 7–9, 191, 216, 233, 295
performance artist (Jon) 245
pets, carp treated as 270
PHCT *see* Pebblebed Heathlands Conservation Trust (PHCT)
Phelan, P. 291
phenomenological perspective 297–8
physical activity, and well-being 13
Pinney, C. 241
place 2–3; and emotion 291–4; and performance art 245; topophilia/topophobia 294–5
place names 36–42, 293–4; and fishermen 265–7; and horse riders 204–7; and the RM 87, 90–1, 123, 293
plagioclimax 134
plants, heathland 131
poet (Barbara Farley) 243–4
politics of nature 151, 289, 290
ponds *37–8*, 51–3
Pope, S. 237–8
population, Pebblebed heathland (Devon) 15
praxeological approach 9, 178
Probert, G. B. 177
Protected Areas (PAs) 127, 128–9
public; and the RM 117; visitor surveys 165–74
public access 77–82; and fencing 71–2
public meetings, on the golf courses 223
purity in nature 289

quarrying 17–23, 30, 152–62, 247; Isle of Harris (Western Isles, Scotland) 161
quasi-objects 6

Rapport, N. 19–20, 232
rationality, and environmental policies 11–12

'ready alert' 93–4
reason 8
recruits, and the Royal Marine training programme 86–7, 116–17
reference points, Royal Marine 87, 90–1
representation of research 297–8
representations of landscape 4–5
riding *see* cyclists/cycling; horse riding
rights, grazing 67, 74–5
Robertson, R. 214
Rosen, P. 178
Royal Marines (RM) 14, 17–23, 30, 84–6, 123–88; comparison to the Squabmoor fishermen 269; and conservation 116–21; and cyclists 190–1; and the heathlands painting *242–52*; and horse riders 209; and memory 121–4; and model aircraft flyers 284–5; training 74, 75; trench digging 49–51, 104; and walkers 225–6; and Williams 32; *see also* training programme of the Royal Marines
Royal Society for the Protection of Birds (RSPB) 25–6, 28; consultation report 70–1; and landscaping after quarrying 155; management of heathland 31–2; and the RM 119; volunteers 132–3, 140, 141–2, 148–9; *see also* Aylesbeare and Harpford Commons
Ruggie, J. 129
Rycroft, S. 4

sacred, the 11–12
Sands, R. R. 176
Schacter, R. 235
Schneider, A. 240, 241
scraping, top soil 50–4, 58–66
Second World War; Dalditch camp 39, 50, 63, 87, 222; and grazing 67
Second World War camp *see* Dalditch camp
sensorimotor practices 100
sensory experience, and the Royal Marine training 115–16
Sheets-Johnstone, Maxine 216
Shinrin-yoku (forest-air bathing) 13
Sid Valley Cycling Club, guide 78–9
Sites of Special Scientific Interest (SSSI) 26, 68
Solnit, R. 213
southern damselfly (*Coenagrion mercuriale*) 26, 46
spatio-temporal relationships 9
Special Area of Conservation (SAC) 26
Special Protection Area (SPA) 26
spiders 141
sport; alternative 177; and anthropology 176–7
Squabmoor reservoir 262–5, *266, 272*
Sunderland, Tom 165

surveys; visitor 165–75; wildlife 137, 140
swaling 64–7, 136–7
SWOT analysis 69–70, 73–4

Tam, K-P. 128
tasks/tools of the volunteers 134–7
taskscapes 10; *see also* pebblebed heathland (Devon)
Taylor, Toby 30–2, 33, *40–1*, 50–1, 61, 165
technologies 8–9
Thomas, N. 241
Tilley, Chris 61, 165, 178–9, 215, 253, 292
time, and the heathland 48
Time Bank 133
tools, volunteer 134–7
topophilia/topophobia 294–5
tourism 14
training programme of the Royal Marines 86–7, *88–96*; endurance course 89, *106*, 107–16; and the landscape 103–6; looking and seeing 94–9; mind and body 101–3; movement 92–4; trench digging by the RM 49–51, 104
Trenchard, Priscilla 241, 251–61; 'Woven Flame' 256, *257–8*
Tuan, Yi-Fu 294–5

Underhill-Day report 69–70

Varley, John 26–9, 48
Venn Ottery Common 25; quarry 156–62
Vergunst, Jo Lee, *Ways of Walking* 215
Vilaça, A. 7
Vivanco, L. 176
volunteers 128, 130, 131–7; feelings about volunteering 142–4, 150–1; relationship with the heathland 144–9; tasks/tools 134–7

Walk On exhibition 239
walkers/walking 17–23, 213–15, 232–3; and anthropology 215–16; and art 236–9; and the character of the landscape 219; comparison to cycling 192–3; and dogs 226–8; and the dwelling perspective 216; as educational 220; and the hunt 219–20; interviewees 217–18; led walks 220–1; maps 229, *230–1*; and memories 223–5; motivations for 216–17; and other users 225–6; physicality of 218–19; views of change in the landscape 222
Walking Artists network 236
Warnier, J-P. 9, 178
'Watching for Nightjars' (Farley) (poem) 243–5
Ways of Walking (Ingold and Vergunst, eds) 215
weapons, and Royal Marine training 91, 99–100
weather 122
Weiner, J. 36
well-being 3, 13
wet and dry routines 103, 110
White, Gilbert 65–6; *Natural History of Selborne* 64
Whitton, Cressida 60–1, 62, 63, 165
wildlife 130–1; and cyclists 190–1; East Budleigh Common 42; and horse riders 207; RM effects on 118–19; spiders 141; visitor awareness of 170–5
Williams, Bungy 32–3, 49–50, 61, 165; lecture for recruits 116–17; map *41–9*; on scraping 54, 62; on swaling 65; on walking tracks 78; on Woodbury Castle 200
Wittgenstein, L. 234
women, walkers 228–9
Woodbury Castle (Iron Age hillfort) 47, 174, 225; and cyclists 199–200, 222; and the performance artist 245; and the RM 89, 90–1
Woodbury Common 39–40, *52–3, 55–7, 67–79*; and the RM 87–90, 124–88; *see also* training programme of the Royal Marines
Woodward, R. 122
Wordsworth, William 214
'Woven Flame' (Trenchard) (basket) 256, *257–8*
Wright, C. 240, 241

Zavestoski, S. 12–13
Zeiske, C. 237

www.ingramcontent.com/pod-product-compliance
Lightning Source LLC
LaVergne TN
LVHW071256150426
836100LV00016B/49